The Michigan Historical Reprint Series

Reprints from the collection of the University of Michigan University Library

The Scholarly Publishing Office
The University of Michigan
University Library

CREMONA'S

TWO TREATISES ON GRAPHICAL STATICS

HUDSON BEARE

London

HENRY FROWDE

OXFORD UNIVERSITY PRESS WAREHOUSE

AMEN CORNER, E.C.

GRAPHICAL STATICS

TWO TREATISES

ON THE

GRAPHICAL CALCULUS

AND

RECIPROCAL FIGURES IN GRAPHICAL STATICS

BY

LUIGI CREMONA

LL.D. EDIN., FOR. MEMB. R.S. LOND., HON. F.R.S. EDIN.
HON. MEMB. CAMB. PHIL. SOC.
PROFESSOR OF MATHEMATICS IN THE UNIVERSITY OF ROME

TRANSLATED BY

THOMAS HUDSON BEARE

B.SC. LOND., ASSOC. M. INST. C.E., F.R.S. EDIN.
PROFESSOR OF ENGINEERING AND APPLIED MECHANICS, HERIOT-WATT COLLEGE, EDINBURGH

Oxford

AT THE CLARENDON PRESS

1890

TRANSLATOR'S PREFACE.

For some years I had used a rough English manuscript summary of Professor Cremona's works on the *Graphical Calculus* and *Reciprocal Figures*, while reading with engineering students of University College, London. As English versions were much wanted, I was advised by Professors Pearson and Kennedy to ask the consent of Professor Cremona to my undertaking their translation, and at the same time they supported my application to the Delegates of the Clarendon Press that they should become the publishers. To both applications a most cordial consent was given; and I take the opportunity of thanking both the Author and the Delegates for the trust they have reposed in me. The translations have been revised by Professor Cremona and certain portions (in particular Chap. I. of *Reciprocal Figures*) have been entirely written by him for the present English edition. I regret that a long delay has occurred in the appearance of this book, due chiefly to pressure of work both on the part of myself and Professor Cremona.

I feel sure that the translation will supply a long-felt want, and be found extremely useful by students of engineering and the allied sciences, especially by those whose work compels them to pay attention to graphical methods of solving problems connected with bridges, roofs, and structures presenting similar conditions.

THE TRANSLATOR.

Heriot-Watt College, Edinburgh.

CONTENTS.

ELEMENTS OF THE GRAPHICAL CALCULUS.

CHAPTER I.

THE USE OF SIGNS IN GEOMETRY.

CHAPTER II.

GRAPHICAL ADDITION.

CHAPTER III.

GRAPHICAL MULTIPLICATION.

CHAPTER IV.

POWERS.

CHAPTER V.

EXTRACTION OF ROOTS.

CHAPTER VI.

SOLUTION OF NUMERICAL EQUATIONS.

CHAPTER VII.

REDUCTION OF PLANE FIGURES.

CHAPTER VIII.

CENTROIDS.

CHAPTER IX.

RECTIFICATION OF CIRCULAR ARCS.

RECIPROCAL FIGURES IN GRAPHICAL STATICS.

ELEMENTS OF THE GRAPHICAL CALCULUS

AUTHOR'S PREFACE

TO THE ENGLISH EDITION.

A GREAT many of the propositions, which form the Graphical Calculus of the present day, have been known for a long time; but they were dispersed in various geometrical works. We are indebted to CULMANN for collecting and placing them at the head of his Graphical Statics; a branch of science, created by him, which is such a powerful help in engineering problems.

The first chapter of this small work, which now appears in English, treats of the use of signs in Geometry, as MOEBIUS conceived them. The succeeding chapters, on Graphical Addition and other arithmetical operations, contain chiefly the graphical calculation of a system of forces in a plane when they are represented by rectilinear segments. The research on centroids, to which the reduction of plane figures serves as an introduction, refers equally to the same subject, being nothing else but the determining of the centres of systems of parallel forces. A special chapter is dedicated to LILL's method of graphical resolution of numerical equations.

As MR. BEARE expressed a wish to translate my little treatise *Il Calcolo Grafico*, and also *Le Figure Reciproche nella Statica Grafica*, for the use of English students, and as the Clarendon Press authorities kindly agreed to publish them, I have been happy to give my consent, as I gave it, some time ago, to Mr. LEUDESDORF for the translation of my *Geometria projettiva*. Whilst reading the translation I have profited by the opportunity to revise the text, and to introduce some improvements.

I take the opportunity of thanking both the Translator, and the Delegates of the Clarendon Press.

THE AUTHOR.

ROME, *July* 1888.

ELEMENTS OF THE GRAPHICAL CALCULUS.

CHAPTER I.

THE USE OF SIGNS IN GEOMETRY.

1. LET O, A, X be three points in a given straight line (Fig. 1), of which O and A are fixed points whilst X moves from O in the direction OA. Further let the *segments* (limited portions of the straight line) OA, OX contain a, x linear units respectively*. Then as long as X remains between O and A, we have $x < a$; when X coincides with A, $x = a$; and as soon as X has passed beyond A, we shall have $x > a$.

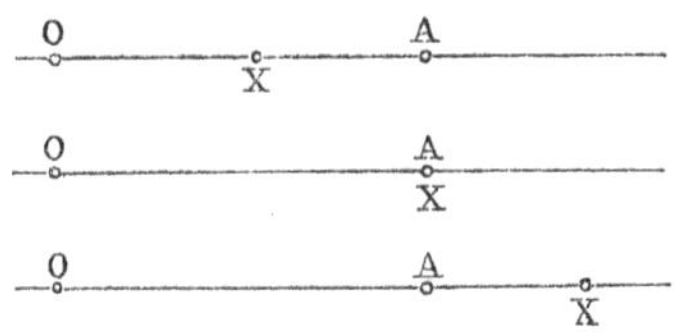

Fig. 1.

If the point X instead of moving from O towards A, were to travel in the opposite direction (Fig. 2), the number x of linear units contained in the segment OX would be considered negative, the number a remaining positive. For example, if X and A were equally distant from O, we should have $x = -a$.

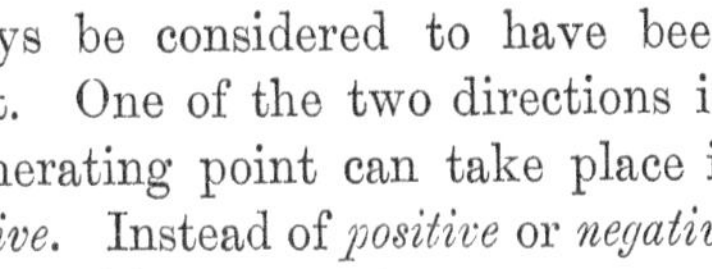

Fig. 2.

A straight line will always be considered to have been described by a moving point. One of the two directions in which the motion of the generating point can take place is called *positive*, the other *negative*. Instead of *positive* or *negative direction* we may also speak of *positive* or *negative sense.*

When a segment of a straight line is designated by the

* The linear unit is supposed to be a segment of unit length measured in the same direction as OA.

number (x) of linear units it contains, its sense is shown by the sign + or — of the number x.

A segment may also be designated by means of the two letters which stand at its ends; for example AB (Fig. 3). In this case we agree to write AB or BA, according as the generating point is conceived to move from A to B, or in the opposite sense. In accordance with this convention, the symbols AB, BA denote two equal magnitudes of opposite* sense, hence the identity

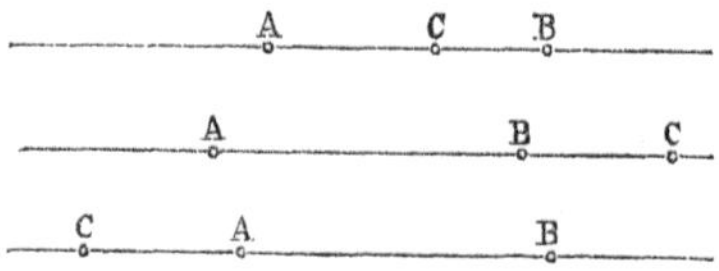

Fig. 3.

$$AB + BA = 0,$$

or $$AB = -BA,\ BA = -AB.$$

Of the two points A, B, the *extremities* of the segment AB, the one A is called the *initial*, and the other B the *final point* of the segment. On the other hand for the segment BA, B is the initial point, and A the final point.

2. Let A, B, C be three points in a straight line. If C lies between A and B (Fig. 3), then

$$AB = AC + CB,$$

and therefore $$-CB - AC + AB = 0,$$

or, since [Art. 1] $-CB = BC$, and $-AC = CA$,

$$BC + CA + AB = 0.$$

If C lies on the prolongation of AB, then

$$AB + BC = AC,$$

hence $$BC - AC + AB = 0,$$

and therefore $$BC + CA + AB = 0.$$

And, finally, if C lies on the prolongation of BA,

$$CA + AB = CB,$$

hence $$-CB + CA + AB = 0,$$

or $$BC + CA + AB = 0.$$

We therefore conclude that †:

If A, B, C are three points (in any order whatever) in a straight line, the identity

$$BC + CA + AB = 0$$

always holds.

* That is to say, two magnitudes of equal arithmetical values, but with opposite algebraical signs, such as $+a$ and $-a$.

† MÖBIUS, *Barycentrischer Calcul* (Leipzig, 1827), § 1, Gesammelte Werke, Bd. 1.

3. From this proposition we obtain an expression for the distance between two points A, B in terms of the distances of these points from a third point O collinear with them which we choose as the *initial point* of the segments. In fact, since O, A, B are three points in a straight line, we have

$$OA + AB + BO = 0,$$

therefore $$AB = OB - OA,$$

or $$AB = AO + OB.$$

4. If $A, B, C, \ldots, M, N$ are n points in a straight line, and if the theorem expressed by the equation

$$AB + BC + \ldots + MN + NA = 0$$

is true for them; then the same theorem is true for $n+1$ points. For if O is another point of the same straight line, then since between the three points N, A, O there exists the relation

$$NA = NO + OA,$$

the above assumed equation becomes

$$AB + BC + \ldots + NO + OA = 0. \qquad \text{Q. E. D.}$$

Now it has already been proved (Article 2) that the theorem is true for $n = 3$, therefore it is also true for $n = 4$, and so on.

5. The sign of a segment AB is undetermined, unless a positive segment of the same straight line has already been given; the direction of this latter segment is called the *positive direction of the straight line.*

For two different straight lines the positive direction of the one is in general independent of that of the other. But if the two straight lines are parallel, we can compare their directions and say that they have the same positive direction when, after having displaced the one line parallel to itself until it coincides with the other, the two directions are found to be identical.

Hence it follows, that two parallel segments AB, CD have the same or opposite signs, according as the direction from A to B coincides with the direction from C to D, or not. If, for example, $ABCD$ is a parallelogram, then

$$AB + CD = 0, \text{ and } BC + DA = 0.$$

If we draw through n given points of a plane $A_1, A_2, \ldots, A_n$, segments A_1A_1', A_2A_2', …, A_nA_n' all parallel to some given direction in the plane until they intersect a fixed straight line $A_1'A_2' \ldots A_n'$, then the sense of one segment determines that of all the others. Two segments A_rA_r', A_sA_s' have the same or opposite sense, according as the points A_r, A_s lie on the same or opposite side of the given straight line $A_1'A_2' \dots A_n'$.

Two equal parallel segments, with the same sign, are called *equipollent*, after Bellavitis.

6. *If A, B, C, D are four collinear points, we have the identity*

$$AD.BC + BD.CA + CD.AB = 0.$$

For the segments BC, CA, AB can be expressed as follows,

$$BC = BD - CD,$$
$$CA = CD - AD,$$
$$AB = AD - BD;$$

now multiply these three equations by AD, BD, and CD respectively and add the results, the right-hand side vanishes, and we obtain the identity we wished to prove.

7. Let p, q, r be three straight lines intersecting in the point O (Fig. 4). Through any point M of the plane draw a

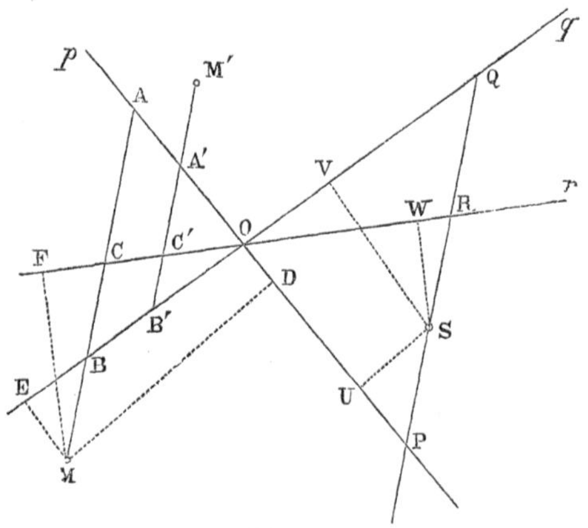

Fig. 4.

transversal, cutting p, q, r in A, B, and C respectively; then from the proposition just proved, we have

$$MA.BC + MB.CA + MC.AB = 0.$$

Now draw, parallel to the transversal ABC, a straight line cutting p, q, r in the points P, Q, R; then the segments BC, CA, AB are proportional to the segments QR, RP, and PQ respectively, and the above equation may therefore be written

$$MA.QR + MB.RP + MC.PQ = 0.$$

If we now draw through any other point M' a new transversal in the fixed direction PQR, cutting p, q, r in A', B', and C', we have similarly

$$M'A'.QR + M'B'.RP + M'C'.PQ = 0;$$

that is to say:

If we draw, through any point M, in a given direction, a transversal which cuts three given concurrent straight lines in A, B, C respectively, then the segments MA, MB, MC are connected by the relation

$$a.MA + b.MB + c.MC = 0,$$

where a, b, c are constants.

From the point M let fall perpendiculars MD, ME, MF upon the three straight lines p, q, r; and also from some arbitrarily chosen point S of the line PQR perpendiculars SU, SV, SW upon the same given straight lines. Then since the triangles MAD, SPU are similar we have

$$MA : MD = SP : SU.$$

Therefore $$MA = \frac{SP}{SU} \cdot MD,$$

and similarly $$MB = \frac{SQ}{SV} \cdot ME,$$

$$MC = \frac{SR}{SW} \cdot MF.$$

The equation

$$MA.QR + MB.RP + MC.PQ = 0,$$

may therefore be written

$$MD.\frac{QR.SP}{SU} + ME.\frac{RP.SQ}{SV} + MF.\frac{PQ.SR}{SW} = 0;$$

that is to say:

If we drop from any point M perpendiculars MD, ME, MF upon three concurrent straight lines, the following relation holds

$$\alpha.MD + \beta.ME + \gamma.MF = 0,$$

where α, β, γ are constants.

The lines MD, ME, MF, instead of being perpendicular to the given straight lines, may be inclined to them at any the same arbitrarily chosen angle; we should then obtain a relation of similar form, by merely altering the values of the constants α, β, γ; the proof however remains the same.

The proof does not necessarily presuppose that the intersection of the three straight lines p, q, r lies at a finite distance;

the proposition is therefore true even if the three given straight lines are all parallel to one another.

8. A plane has two sides which face the two regions into which it divides space. Let a perpendicular be drawn through any point O of the plane, and let the positive direction of this perpendicular be fixed. If Ol be any positive segment of this straight line, then the region in which l lies is called the positive region, and the side which looks toward l is called the positive face.

Now let an observer, standing with his feet at O, and his head at l observe a rotational motion in the plane (Fig. 5); this can take place in two senses, either from left to right [dex-

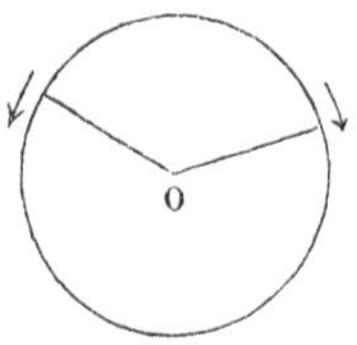

Fig. 5.

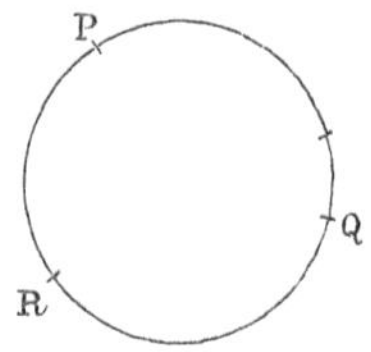

Fig. 6.

trorsum, in the sense of rotation of the hands of a watch], or from right to left [sinistrorsum]. The former sense is called *positive*, the latter *negative.*

Let P, Q, R be three points on a circle in the plane (Fig. 6); the points P, Q divide the circumference into two arcs PQ, one of which contains R. If we take as positive the sense in which one of the two arcs has been described, the other arc has negative sense. If we fix the positive arc PQ, then the sense of any arc, and of any rotational motion in the plane will be fixed; and thereby the positive face of the plane is also fixed, as it is the one on which the observer must stand in order that the positive arcs may seem to him to be described in the sense of the motion of the hands of a watch. The *positive sense* of a plane is that of its positive arcs.

9. Let a, b be the positive directions of two straight lines in a plane, intersecting in the point O (Fig. 7), and let OP, OQ be two positive segments of these straight lines, each of length equal to unity. By the angle ab between these two lines, we mean the circular arc PQ described in the positive sense of the plane. In order that the angle may be fixed it is necessary to fix both the positive directions of the two straight lines and the positive sense of the plane; but we may add to any angle any number of complete rotations either positive or negative, i.e. (if n is an integer),

$$ab \pm 360^\circ \times n = ab.$$

If OA, OB are two positive segments of the straight lines a, b, the angle ab can also be denoted by $\widehat{OA.OB}$, or more briefly by AOB.

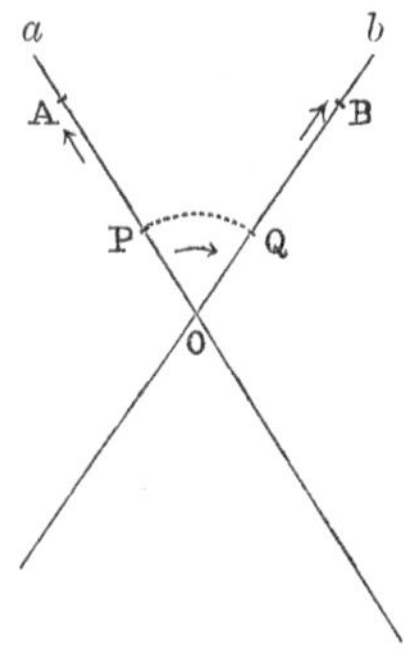

Fig. 7.

The sum of the angles ab, ba is equal to any number of complete revolutions; we may therefore write

$$ab + ba = 0,$$

or $$ba = -ab, \text{ or } ab = -ba.$$

that is to say, ab and ba can be regarded as of equal magnitude and opposite sense *.

This leads us to consider the positive rotation ab as equivalent to the negative rotation $-ba$; or in other words, the angle ab is the circular arc PQ described in the positive sense of the plane, or the circular arc QP described in the negative sense and then taken with the $-$ sign: $PQ = -QP$.

A negative angle is one described by a negative rotation, or by negative arcs.

Analogously we have

$$AOB + BOA = 0;$$

that is, AOB, BOA are two angles of equal magnitude and opposite sense.

10. Let the directions a, b, c of three straight lines in the plane be given, and suppose them to be drawn from the same point O, and to be extended on only one side of it, for the angle between two straight lines is independent of their absolute position. Then if in turning round O in the positive sense of the plane, we meet with the three straight lines in the order acb (Fig. 8), we have the identity

$$ca = cb + ba,$$

hence $$-cb + ca - ba = 0.$$

But $$-cb = bc, \quad -ba = ab,$$

and therefore $$bc + ca + ab = 0.$$

* Baltzer, *Analy. Geometrie*, § 9.

If the order of the succession is abc (Fig. 9), then

$$bc + ca = ba,$$

or $$bc + ca - ba = 0,$$

and therefore $$bc + ca + ab = 0.$$

Accordingly we have this proposition:

If a, b, c are three straight lines in the same plane, in any order whatever, the identity

$$bc + ca + ab = 0$$

is always true.

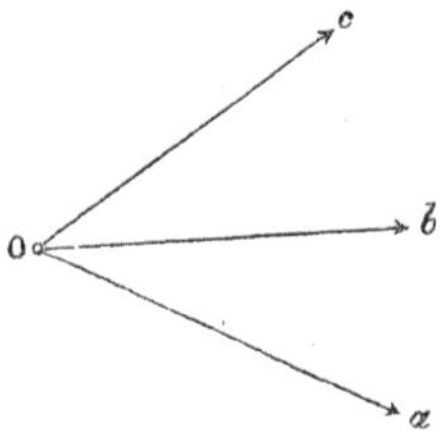

Fig. 8.

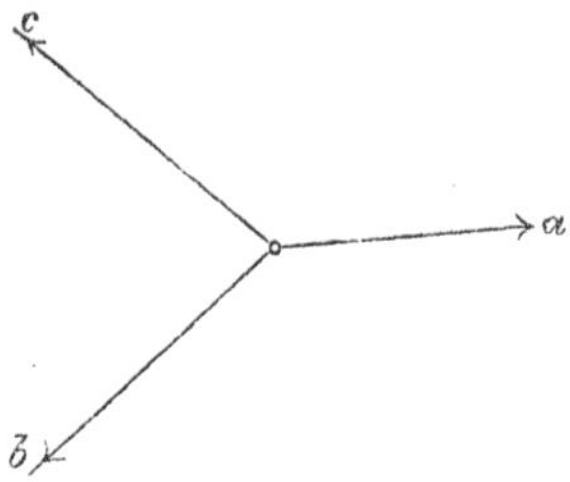

Fig. 9.

11. From this we obtain, by a procedure similar to that for segments (Art. 3), an expression for the angle between two straight lines a, b, in terms of the angles, which they make with a third straight line o, taken anywhere at pleasure in the given plane. In fact if o, a, b are directions in one and the same plane, we have

$$oa + ab + bo = 0,$$

therefore $$ab = ob - oa,$$

or $$ab = ao + ob.$$

12. Three points A, B, C which do not lie in one straight line, are the vertices of a triangle (Fig. 10). Let us consider that we pass round its periphery continuously, that is, passing through each point once and through no point more than once: then each vertex is the final point of one side and the initial point of the following side. This can be done in two ways, that is to say in two opposite directions; namely in the sense ABC or in the sense ACB.

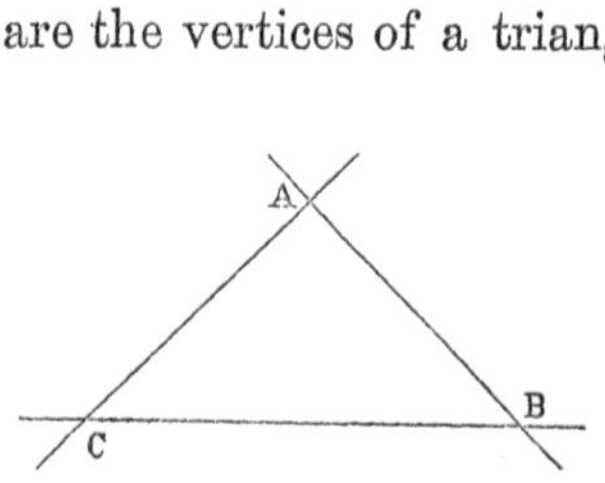

Fig. 10.

The sense BCA or CAB does not differ from ABC, and similarly neither CBA nor BAC is different from ACB.

The area of the triangle lies to the right or to the left hand, according as we go round the periphery in the positive or negative sense; for this reason we consider the areas ABC, ACB as equal but opposite: the first as positive, the second negative. We may suppose the area ABC (or ACB) to be described by a revolving line of variable length, of which one end is fixed at A, whilst the other describes the segment BC (or CB). Now this rotation takes place in the positive (or negative) sense of the plane; for this reason also we consider the area as positive (or negative)*.

The necessary and sufficient condition that three points A, B, C may lie in one straight line, is that the area ABC is zero.

13. PROPOSITION. If O *is any point whatever, in the plane of the triangle* ABC (Fig. 11), *we always have the identity*

$$OBC + OCA + OAB = ABC\ †.$$

Proof. If O lies within the triangle ABC, then of course the latter is the sum of the triangles OBC, OCA, OAB.

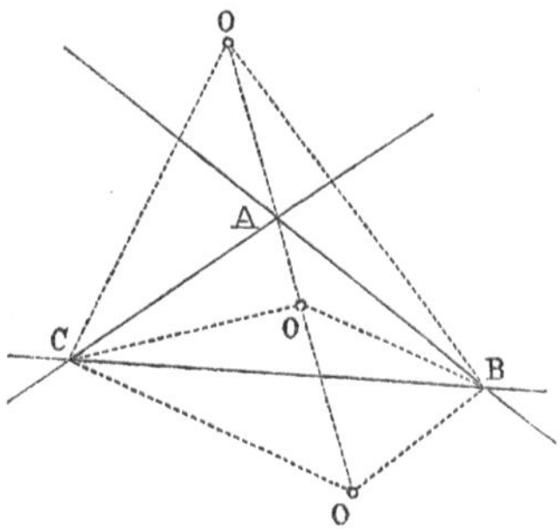

Fig. 11.

If O lies within the angle BAC, but upon the other side of BC, we have

$$OCA + OAB - OCB = ABC\ ;$$

but $$OCB = -OBC,$$

therefore $$OBC + OCA + OAB = ABC.$$

Finally, if O lies within the opposite vertex of BAC, we have

$$OBC - OAC - OBA = ABC,$$

and hence $$OBC + OCA + OAB = ABC.$$ Q. E. D.

It follows from the remark at the end of Art. 12, that if A, B, C are three points in a straight line, then wherever O may be we have

$$OBC + OCA + OAB = 0.$$

14. It follows from this proposition, that the area of the triangle ABC may be regarded as generated by the motion of a revolving line of variable length (*radius vector*), of which one end is fixed at O (the *pole*), whilst the other describes the

* MÖBIUS, loc. cit. § 17. † Ibid. § 18.

periphery (outline) in the sense denoted by the given expression ABC.

This remark and the above proposition would remain unaltered, even if BC were no longer a segment of a straight line, but an arc of a curve *.

15. *If O is any point whatever in the plane of the parallelogram $ABCD$* (Fig. 12), *we have*

$$OAB + OCD = \tfrac{1}{2}\, ABCD,$$

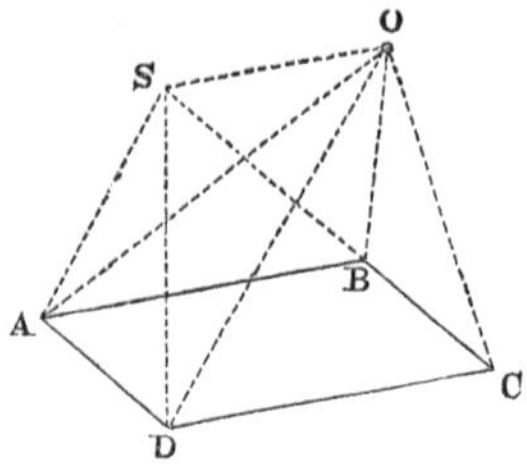

Fig. 12.

For, using S to denote the point, in which the side BC is cut by the straight line, drawn through O parallel to AB, we have (Art. 13),

$$SAB + SBC + SCA = ABC.$$

But $SBC = 0,\ \ SCA = SCD,$

$$SAB = OAB,\ \ SCD = OCD,$$

therefore

$$OAB + OCD = ABC = \tfrac{1}{2}\, ABCD. \qquad \text{Q. E. D.}$$

Since $\frac{1}{2} ABCD = \dot{D}AB$, the above equation may also be written

$$ODC = OAB - DAB.$$

16. Let (Fig. 13) p, q, r be three straight lines, which form a triangle ABC; and let O and M be two points in its plane, of which the first is considered as fixed or given, and the other as variable. Draw from the points O and M to the straight line p in any direction the two parallels OU, MD, and similarly to q the parallels OV, ME, and to r the parallels OW, MF, also in any directions whatever.

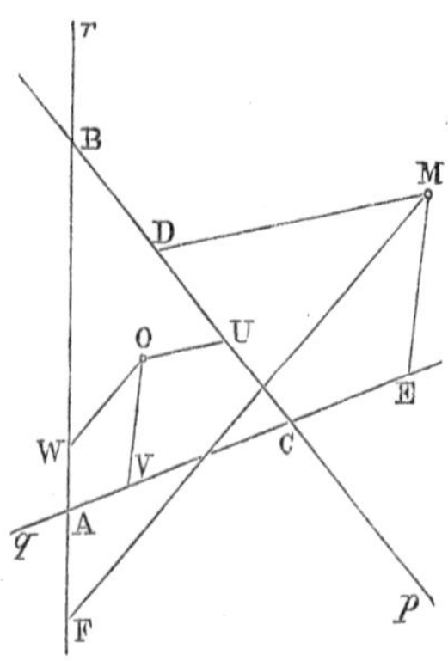

Fig. 13.

The areas of the triangles OBC, MBC are proportional to the distances of their vertices O, M from the common base BC, and therefore also to the segments OU, MD; hence we have

$$OBC : MBC = OU : MD,$$

* And therefore also, if BC, CA, AB were three arcs, which do not intersect, see Art. 19.

or $$MBC = \frac{OBC}{OU} \cdot MD,$$

and similarly $$MCA = \frac{OCA}{OV} \cdot ME,$$

$$MAB = \frac{OAB}{OW} \cdot MF.$$

But from (Art. 13)

$$MBC + MCA + MAB = ABC,$$

therefore $$\frac{OBC}{OU} \cdot MD + \frac{OCA}{OV} \, . \, ME + \frac{OAB}{OW} \, . \, MF = ABC.$$

If we vary the position of the point M in the plane, whilst keeping the directions OU, OV, OW fixed, then in the above equation only the lengths MD, ME, MF change; we obtain therefore this Theorem:

If we draw in given directions from any point M in the plane of a given triangle, the straight lines MD, ME, MF meeting the sides of this triangle, then these straight lines are connected by the relation

$$(\dagger) \quad \alpha \, . \, MD + \beta \, . \, ME + \gamma \, . \, MF = \delta,$$

the quantities α, β, γ, δ being constants.

The proposition is still true if two of the three given straight lines p, q, r are parallel to one another. For example, let q, r be parallel, and let us draw a straight line s, which is parallel neither to q, r, nor to p. If now we draw through any point M, in directions chosen at pleasure, the straight lines MD, ME, MF, MG to the straight lines p, q, r, s, then from the proposition just proved, since p, q, s form a triangle, MD, ME, MG are related by an equation of the form ($\dagger$), which may be written thus

$$\alpha \, . \, MD + \beta \, . \, ME + MG = \delta\,;$$

and similarly since p, r, s form a triangle, we shall obtain a relation of the same form

$$\alpha' \, . \, MD + \gamma MF + MG = \delta'.$$

between MD, MF, MG.

Subtracting this equation from the foregoing one, we have

$$(\alpha - \alpha') \, MD + \beta \, . \, ME - \gamma \, . \, MF = \delta - \delta',$$

that is to say, MD, ME, MF are also connected by a relation of the form ($\dagger$). Q. E. D.

This proposition is a generalisation of the one (in Art. 7)

concerning three straight lines p, q, r which intersect in a point situated at either a finite or infinite distance. In the special case mentioned the constant δ is zero.

17. We shall call that line a *circuit* which a point describes whilst it moves in a plane from one position (the initial) to another position (the final) continuously, that is without ever leaving the plane. The circuit is closed if the final position coincides with the initial position; it is open if this is not the case. If the circuit intersects itself, we call the points of intersection nodes, and the circuit a *self-cutting one*.

If the circuit is formed of rectilinear segments, it is said to be *polygonal*, or simply a *polygon*.

Any circuit can be described, like the periphery of a triangle (Art. 12), in two opposite senses. In order that the sense of a circuit may be fixed, it is sufficient to know the order of succession of *two* points of it, if the circuit is open, and of *three*, if it is closed.

18. A closed circuit without nodes encloses within itself an *internal* finite region of the plane, and divides it from the rest of the plane, which is *external* and infinite. The *area* bounded by the circuit is the measure of the interior region, and it is considered to be *positive* or *negative*, according as it lies to the right or left of an observer on the plane, who passes along the circuit in the given sense.

19. PROPOSITION. If $ABCD...MNA$ (Fig. 14) *is any closed circuit, and O a point in its plane, then the sum of all the triangles (or sectors),*

$$\Sigma = OAB + OBC + OCD + \ldots + OMN + ONA,$$

is a constant quantity for any position whatever of the pole O.*

Proof. Let O' be another point in the plane; then from the proposition in Art. 13,

$$\begin{aligned} O'AB &= OAB + OBO' + OO'A, \\ O'BC &= OBC + OCO' + OO'B, \\ O'CD &= OCD + ODO' + OO'C, \\ &\&c. \qquad\qquad \&c. \\ O'MN &= OMN + ONO' + OO'M, \\ O'NA &= ONA + OAO' + OO'N. \end{aligned}$$

* MÖBIUS, *Baryc. Calcul*, § 165, Ges. Werke, Bd. 1; *Statik* (Leipzig, 1837), § 45, Ges. Werke, Bd. 3.

Adding we have,

$$O'AB + O'BC + O'CD + \ldots + O'MN + O'NA$$
$$= OAB + OBC + OCD + \ldots + OMN + ONA = \Sigma,$$

since all the other terms cancel, because they occur in pairs of equal and opposite terms, as, for example, $OO'A$ and OAO', $OO'B$ and OBO', and so on. We may consider the magnitude Σ as

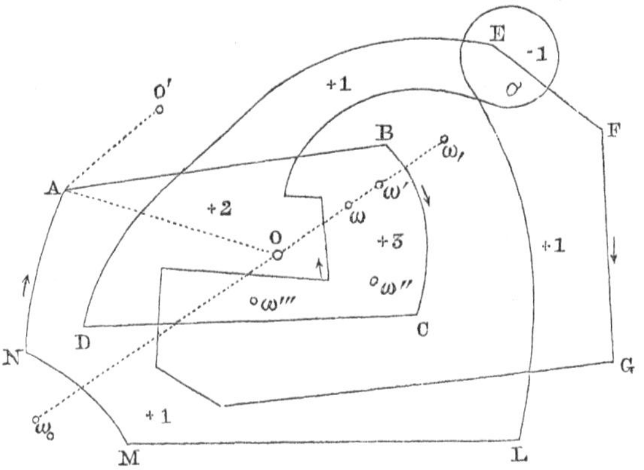

Fig. 14.

generated by the motion of a revolving line OX (radius vector) of variable length, which has one end fixed at the pole O, whilst the other describes the given circuit in the given sense.

20. If a radius vector ΘY be rotated in a given plane about a fixed point Θ, and if it pass over any point O of the given plane, we shall call the *passage*, *positive* or *negative*, according as the radius vector ΘY in passing through O is in the act of describing a positive or negative rotation.

Lemma. If a radius vector ΘY, moveable in a plane about a fixed point Θ, starting from the original position ΘA, describes successively the angles a_1, a_2, &c., &c...., and if after having passed p times positively, and n times negatively, over a given point O it returns to its original position ΘA, then the difference $p-n$ is independent of the order of succession of the angles a.

It will be sufficient to show, that if we interchange a_r, a_{r+1} the difference $p-n$ is unaltered. We are at liberty to suppose that the angles a are less than 180°, because if a_r were greater than 180° we could divide it into parts each less than 180°.

If a_r and a_{r+1} are of the same sign, the radius vector ΘY will either describe the angle $a_r + a_{r+1}$, or the angle $a_{r+1} + a_r$, hence it will pass over the same positions and in the same sense; and therefore neither p nor n will be changed.

Now suppose that a_r and a_{r+1} are of opposite sign. Before the interchange, let us suppose that at the completion of the angles a_{r-1}, a_r, a_{r+1}, the moving radius vector takes respectively the positions ΘY_{r-1}, ΘY_r, ΘY_{r+1} (Fig. 14 *a*), and after the interchange at the completion of the angle a_{r+1} let it take up the position $\Theta Y_r'$. Then, if the point O lies in one of the

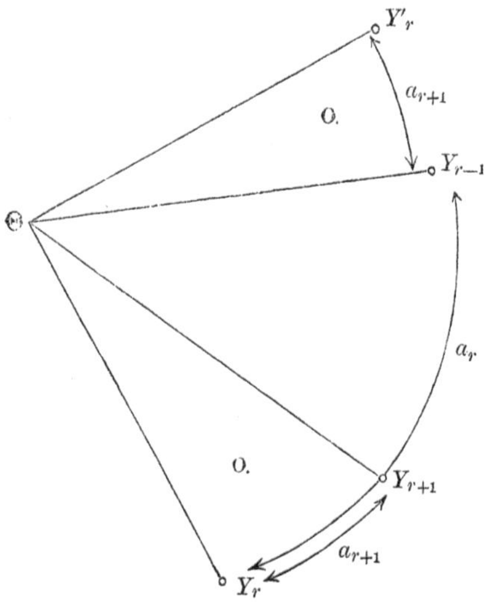

Fig. 14 *a*.

angles $Y_r \Theta Y_{r+1} = Y_{r-1} \Theta Y_r' = a_{r+1}$, the interchange will decrease or increase by unity each of the numbers p, n. If, on the other hand, O lies outside these angles, both these numbers will be unaltered. In every case therefore the difference $p-n$ is unchanged.

Corollary.—The difference $p-n$ is equal to the number (positive or negative) of revolutions contained in the sum $a_1 + a_2 + \ldots$ In fact, let $\kappa . 360 + \gamma$ be the sum of the positive a's, and $-(h\,360 + \gamma')$ the sum of the negative a's. Now γ and γ' are each less than 360°, and as the final position of the *radius vector ΘY is supposed to coincide with its original position ΘA,* we must necessarily have $\gamma = \gamma'$. But by virtue of the preceding lemma the difference $p-n$ will remain unaltered if, instead of describing the angles a_1, a_2, a_3, &c., in succession, we describe the rotation $\gamma-\gamma'+k\,360^\circ-h\,360^\circ$, or the rotation $k\,360^\circ-h\,360^\circ$ (as the equal and opposite angles γ and $-\gamma'$ can be neglected), since this leaves the numbers p, n unchanged, or increases or diminishes each of them by unity. Now, in describing each of the k (or h) positive (negative) rotations, we make a positive (negative) passage through the point O; therefore

$$p-n = k-h.$$

21. THEOREM. Let any given closed circuit whatsoever, in a plane, be described in a given sense by a point X, returning to its original position, after having passed over all the points of the circuit. Take a point O in the plane, and let Σ be the algebraic sum of the sectors described in succession by the radius vector OX. Then the sum Σ remains constant wherever O may be taken *.

Let us imagine the plane divided by a close network of lines into very small areas, which we shall call *elementary areas*, so small that the circuit does not pass through the interior of any one of them, with the exception of those that form part of the contour. If while the point X describes the circuit it happens that the radius vector OX in passing through certain positions changes its sense of rotation, we shall suppose that the straight lines forming these special positions of the radius vector form part of the network. Then it is not possible for any elementary area to be partially described by the radius vector OX, but it will be either totally described or not at all.

Having premised this, then, during the whole movement of the point X in the circuit, let any elementary area whatever ω be described by the radius vector OX, p times positively, n times negatively. Then the area ω will be contained $p-n$ times in Σ, or Σ will be the sum of $(p-n)\,\omega$ extending over all the elementary areas of the plane. It will be therefore sufficient to show that the coefficient $p-n$ does not vary with the pole O.

* DE MORGAN, *Extension of the word area*, (Cambridge and Dublin Math. Journal, vol. v. 1850). For the treatment of this argument the author is indebted to the suggestions of Professor Gabriele Torelli, of Naples.

If we join the point X to a point Θ taken inside the area ω, and if we produce the straight line ΘX beyond Θ, for example, to meet the circuit in Y, then it is evident that every time the radius vector OX describes the area ω in one sense, the straight line ΘY passes through O in the same sense, and conversely. Therefore the number of times OX passes through ω will be equal, in sense and absolute value, to the number of times ΘY passes through O. Therefore if $(k-h)\ 360°$ are the number of complete rotations of the radius vector ΘY, the coefficient of the elementary area ω in the sum Σ will be $k-h$, that is, is independent of the pole O.

22. A given closed and self-cutting circuit (Fig. 15) divides the plane into a definite number of finite spaces S_1, S_2,... contiguous to one another. Each of these is bounded by a circuit without nodes; so that the whole plane consists of

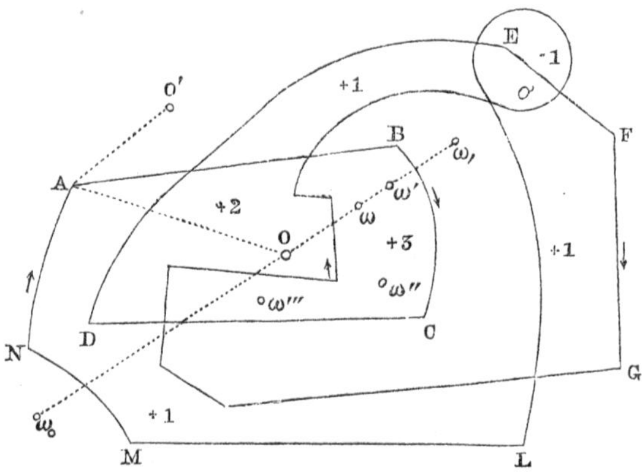

Fig. 15.

these spaces and of the remaining (external) infinite region, which latter we shall denote by S_0.

Let ω and ω' be two elementary areas or elements of the plane, which can be joined by a straight line that does not cross the circuit, and let us take the pole O upon the continuation of the straight line $\omega'\omega$. It is evident that the radius vector OX cannot pass over ω' without at the same time passing over ω in the same sense; ω and ω' will therefore enter into Σ with the same coefficient. The elements ω'', ω'''... have also this same coefficient, if the circuit does not pass between

ω' and ω'', or between ω'' and ω''', &c. Since we can thus conjoin all the elements in succession of one and the same space S, therefore all the elements of S will appear in the sum Σ with the same coefficient c. That is to say, S appears in the sum Σ with the coefficient c. If therefore $c_1, c_2, \ldots$ are analogous coefficients for the spaces $S_1, S_2, \ldots$, we have

$$\Sigma = c_1 S_1 + c_2 S_2 + \ldots,$$

if we understand that $S_1, S_2, \ldots$ at the same time express the areas of the spaces represented by these symbols.

Next let ω, ω_1 be two elements, between which the circuit passes once; and let ω lie on the right and ω_1 on the left of the circuit which passes between ω and ω_1, in the given sense. Take the pole O upon the continuation of the straight line $\omega_1 \omega \ldots$; now if X traverses that part of the circuit which lies between ω and ω_1, the radius vector OX will describe ω once with a positive rotation, without describing ω_1, whilst for all other parts of the circuit the elements ω and ω_1 will be described simultaneously in the same sense. The coefficient of ω will therefore exceed that of ω_1 by 1; that is to say, if in passing from one space to a neighbouring one we cross the circuit once from right to left*, then the coefficient of the first space exceeds that of the other by unity.

The infinite region S_0 has the coefficient zero; for if ω_0 is an element, which lies outside the spaces S_1, S_2, &c...., then it is clear, that we can give the pole O such a position, that the (finite) radius OX never passes through ω_0, wherever X may lie on the circuit.

Any space from which we can get to S_0 by crossing the circuit only once, has the coefficient $+1$ or -1, according as the crossing takes place from right to left, or from left to right. In general if we draw from a point in any space S a straight line to a point of S_0, and if this straight line crosses the circuit m times from right to left and n times from left to right, then the coefficient of S is equal to $m-n$.

23. If the circuit has no nodes, we have a single finite space S, and this has the coefficient $+1$ or -1, according as the

* From right to left is always to be taken in the sense of a person describing the circuit in the given sense; the particular sense is indicated in the figure by an arrow.

circuit has been described positively (Fig. 15*a*) or negatively (Fig. 15*b*). In this case therefore we have

$$\Sigma = \pm S,$$

that is to say, If *the circuit is not a self-cutting-one then the sum* Σ *is the area of the space enclosed by the circuit.*

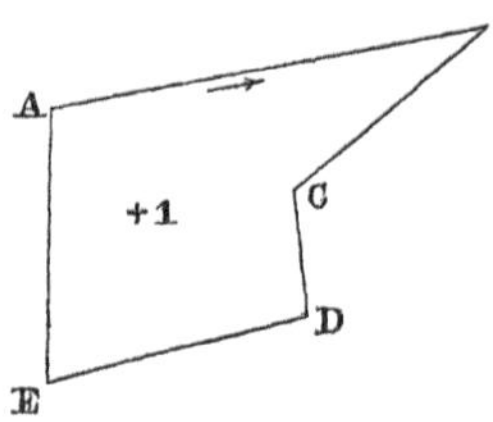

Fig. 15 *a*.

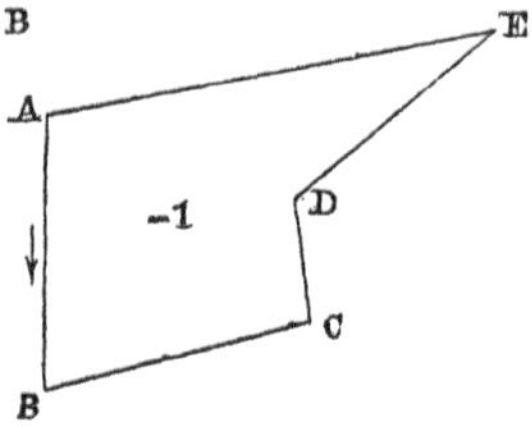

Fig. 15 *b*.

This property naturally leads us to consider the sum Σ as defining the *area* of any self-cutting circuit*.

24. A self-cutting circuit can be decomposed into circuits which are not self-cutting, by separating the (curvilinear)

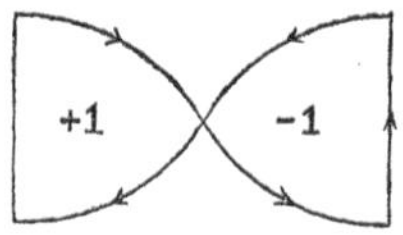

Fig. 16 *a*.

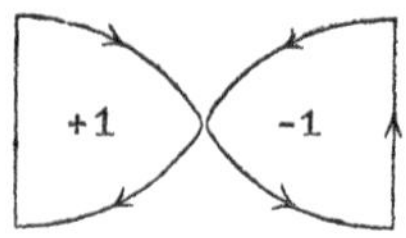

Fig. 16 *b*.

angles, formed by the branches which intersect at each node, without altering at all the sense (i.e. the direction of the arrows)

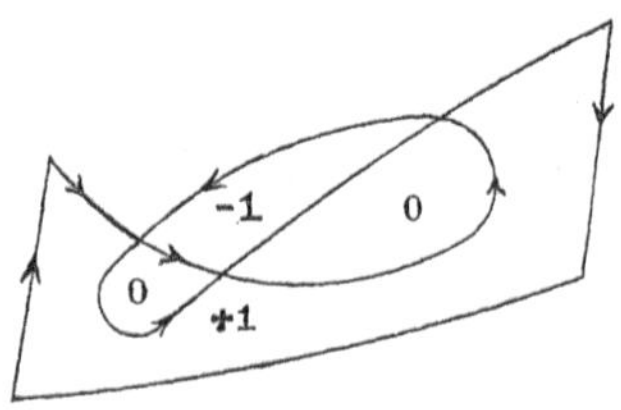

Fig. 17 *a*.

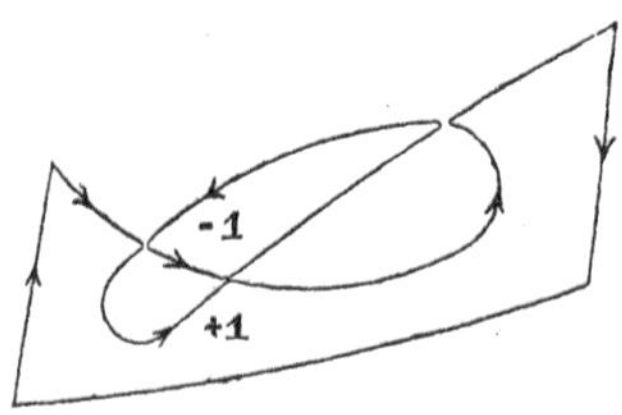

Fig. 17 *b*.

of the branches themselves. Consider, for example, Figs. 16 and 17; in each a self-cutting circuit is resolved into two

* Besides the paper by DE MORGAN previously mentioned, see MÖBIUS, *Ueber die Bestimmung des Inhalts eines Polyeders* [Berichte der Königl. Sächs Gesellsch. der Wissenschaften zu Leipzig, 1865), § 13 and following; Ges. Werke, Bd. 2.]

simple ones; also Fig. 18, where a self-cutting circuit is resolved into four simple ones.

The spaces with negative coefficients are in this way separated from those with positive coefficients; and of two

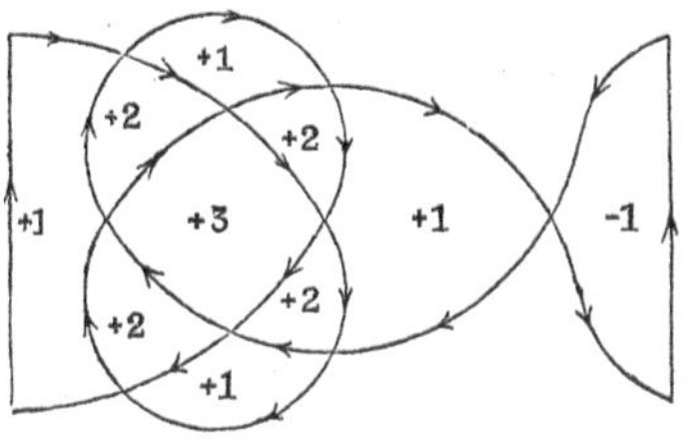

Fig. 18 *a*.

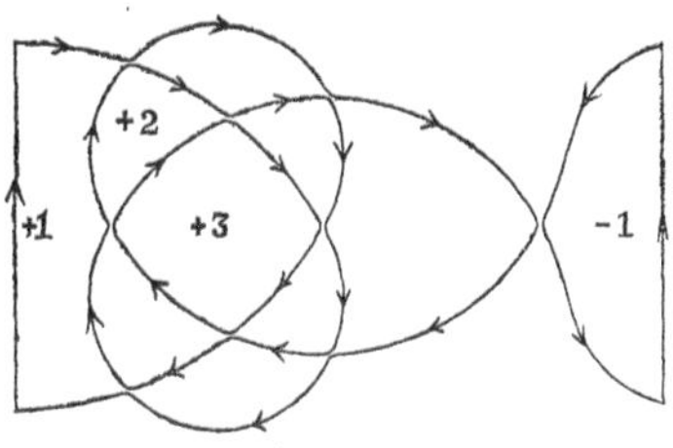

Fig. 18 *b*.

spaces whose coefficients have the same sign, the one whose coefficient is greater in absolute value, lies wholly within the other. Thus, for example (if we denote by S_r the space whose coefficient is r), S_2 is inside S_1; S_3 inside $S_2, \ldots$, S_{-2} inside $S_{-1}, \ldots$. Hence it follows that the area Σ can be expressed as a sum of spaces, which all have positive or negative unity for their coefficient. For this purpose it is sufficient to take the area S_r once for itself, and once more with the area S_{r-1}, within which it lies; that is to say, we sum the spaces S_r and $S_{r-1}+S_r$ instead of $2S_r$ and S_{r-1}, and so on. Consider for example (Fig. 18) where the area is equal to

$$S_3+(S_2+S_3)+(S_1+S_2+S_3)-S_{-1}\text{*}.$$

By the *area* of a system of closed circuits we understand the algebraic sum of the areas of the single circuits. Thus, for

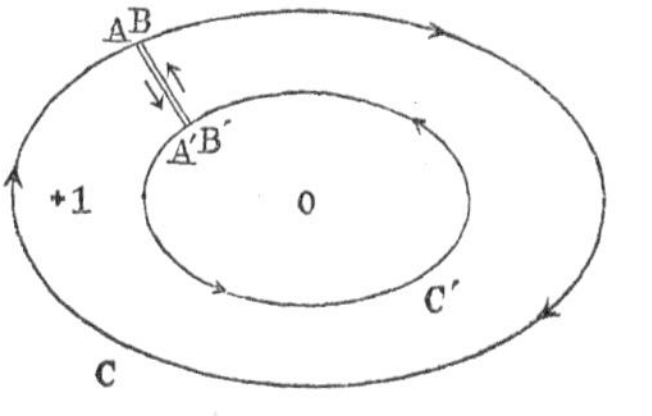

Fig. 19.

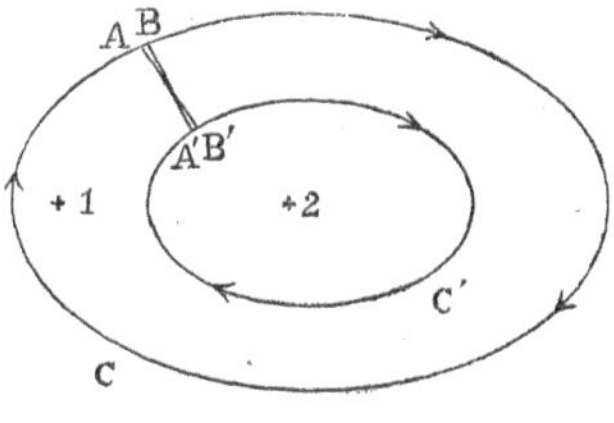

Fig. 20.

example, the ring inclosed between the two oval curves in Fig. 19 is the area of the circuits ABC, $A'C'B'$; on the other hand, the

* CULMANN, *Graphische Statik*, 2d ed. (Zürich, 1875), Nr. 26.

area of the circuits ABC, $A'B'C'$ (of Fig. 20) is equal to that ring plus twice the internal area $A'B'C'$. In both cases we can substitute for the two circuits a single one $AA'C'B'BCA$ (Fig. 19) or $AB'C'A'BCA$ (Fig. 20), where the points B, B' are considered as infinitely near to A, A' respectively. In (Fig. 21) the two

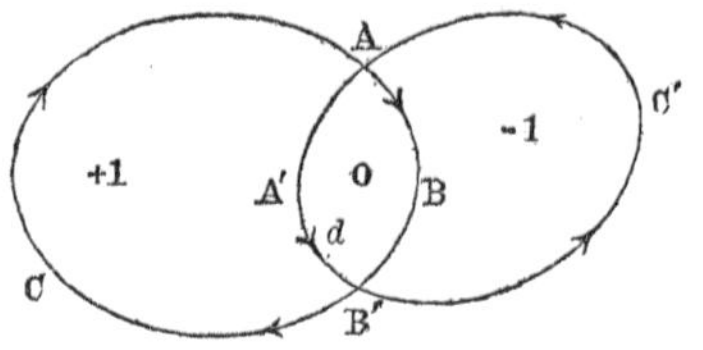

Fig. 21.

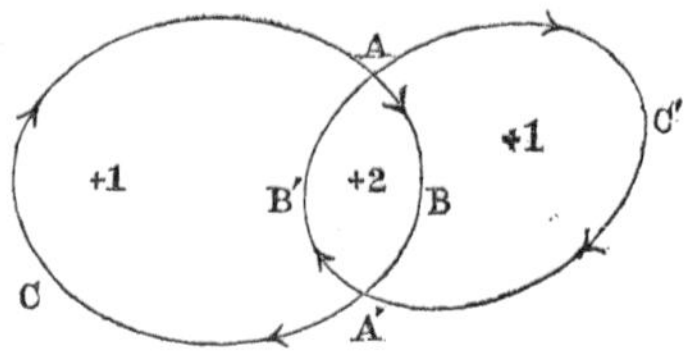

Fig. 22.

circuits intersect; the area of the circuits ABC, $A'B'C'$ is equivalent to that of the circuits $AA'B'C$, $ABB'C'$. In Fig. 22 the area of the circuits ABC, $A'B'C'$ is equivalent to that of the circuits $ABA'B'$, $AC'A'CA$. The two circuits can, in each case, be replaced by a single one.

25. If the two closed polygons $CDE \ldots M$, $C'D'E' \ldots M'$, in a plane, have their sides CD, $C'D'$, DE, $D'E'$, $\ldots$ MC, $M'C'$ respectively equipollent, the sum of the parallelograms $CDD'C'$ $DEE'D' \ldots$ $MCC'M'$ is zero. It will be sufficient to prove this for the case of the triangle CDE.

Taking D as the pole of the contour $CC'D'E'E$, we have from the theorem of Art. 19,

$$DCC' + DC'D' + DD'E' + DE'E + DEC = CC'D'E'E.$$

But the two first triangles together form the parallelogram $DCC'D'$; similarly the third and fourth triangles form the parallelogram $EDD'E'$. Also

$$CC'D'E'E - DEC = CC'D'E'E - D'E'C' = E'ECC',$$

which is a parallelogram.

Wherefore:

$$DCC'D' + EDD'E' + CEE'C' = 0.$$

From this it follows that if $CDE \ldots M$ is a closed polygon whose n sides are the bases of n triangles whose vertices are the points A_1 $A_2 \ldots A_n$, respectively, (which are taken anywhere in the plane of the polygon,) the sum of the triangles

$$A_1CD + A_2DE + \ldots + A_nMC,$$

does not change when the polygon is moved parallel to itself in its plane. In fact, if $C'D'E' \dots M'$ is another polygon, whose sides are equipollent to those of the given one, we have

$$A_1CD = A_1C'D' + C'DD',$$
$$A_2DE = A_2D'E' + D'EE',$$
$$\cdots\cdots\cdots$$
$$A_nMC = A_nM'C' + M'CC';$$

summing up we have,

$$A_1CD + A_2DE + \dots A_nMC = A_1C'D' + A_2D'E' + \dots,$$

because, as we have shown above, the sum

$$2\,[C'DD' + D'EE' + \dots M'CC']$$

is equal to zero.

26. THEOREM. *If the rectilinear segments A_1B_1, A_2B_2, $A_3B_3, \dots A_nB_n$ of given magnitude and position in a plane are equipollent to the sides of a polygon (i.e. of any rectilinear closed circuit, whether self-cutting or not), then the sum of the triangles*

$$OA_1B_1 + OA_2B_2 + OA_3B_3 + \dots + OA_nB_n,$$

is constant wherever the pole O may be taken, at a finite distance. But if the given segments are not equipollent to the sides of a closed polygon, then this sum is not constant except for such points O, as are equidistant from a fixed straight line.*

Proof. Construct the crooked line $CDE \dots MN$, of which the successive sides $CD, DE, \dots MN$ are respectively equipollent to the given segments A_1B_1, A_2B_2, $A_3B_3, \dots A_nB_n$; so that the figures A_1B_1DC, $A_2B_2ED, \dots, A_nB_nNM$ are parallelograms.

Then from (Art. 15),

$$OA_1B_1 = OCD - A_1CD,$$
$$OA_2B_2 = ODE - A_2DE,$$
$$\&c. \qquad \&c.$$
$$OA_nB_n = OMN - A_nMN,$$

and also from the proposition in Art. 19

$$OCD + ODE + \dots OMN + ONC = CDE \dots MNC,$$

hence by addition we have

$$OA_1B_1 + OA_2B_2 + \dots + OA_nB_n = CDE \dots MNC + OCN - (A_1CD + A_2DE + \dots + A_nMN).$$

If the given equipollent segments form a closed polygon, that is, if the point N coincides with C, then the area of OCN is zero, provided that the point O remains at a finite distance, and therefore the sum $OA_1B_1 + OA_2B_2 + \dots + OA_nB_n$

* APOLLONIUS, *Loci Plani*, lib. 1. L'HUILIER, *Polygonométrie*, 1789, p. 92. MÖBIUS, *Statik*, § 46.

has a value independent of the position of O. Hence it follows that, in the special case where CN is zero, the above-mentioned sum either has the value zero for every point O of the plane, or else it vanishes for no single point O (lying at a finite distance).

If N does not coincide with C, the above sum will remain unaltered, so long as the area of the triangle OCN does not alter; that is, so long as the point O remains at the same distance from the straight line CN.

If we change this distance, and take a new pole O', we shall have

$$O'A_1B_1 + O'A_2B_2 + \dots + O'A_nB_n = CDE \dots NC + O'CN$$
$$-(A_1CD + A_2DE + \dots + A_nMN).$$

Take the pole O' at such a distance from CN, that the area of the triangle $O'CN$ is equal to

$$A_1CD + A_2DE + \dots + A_nMN - CDE \dots NC,$$

then the sum $O'A_1B_1 + O'A_2B_2 + \dots + O'A_nB_n = 0$.

The straight line (parallel to CN), which is the locus of those points O' for which this sum is zero, we call r. If we take the point C, i.e. the arbitrary initial point of the crooked line $CDE \dots$, upon r, then the area of $O'CN$ is zero, and therefore the sum of the triangles

$$A_1CD + A_2DE + \dots + A_nMN$$

is equal to the area $CDE \dots MNC$. If we keep to this choice of C, i.e. if we agree that C shall be a point in the line r, then for any point whatever O we shall have

$$OA_1B_1 + OA_2B_2 + \dots = OCN.$$

27. Conversely, if the sum OA_1B_1, OA_2B_2, &c. ... is the same for every point O in the plane, the segments A_1B_1, A_2B_2, &c. ... are equipollent to the sides of a closed polygon. If there are two segments, they will therefore be parallel, equal, and opposite in sense. If we take the point O on one of them, we see that the sum is half that of the parallelogram formed by the two segments.

28. In the special case, where all the given segments meet in a common point C, the sum of the triangles

$$A_1CD + A_2DE + \dots + A_nMN, \text{ or else } CDE + \dots + CMN$$

is identical with the area of the polygon $CDE \dots NC$ (Art. 23); and therefore the common point C must also be a point in the straight line r. This is tantamount to saying that in this case

r coincides with the straight line CN, which joins the extremities of the crooked line $CDE \ldots MN$.

The same conclusion holds good if the given segments lie upon straight lines, which all intersect in the same point C; since we can substitute for the triangle OA_rB_r the triangle OCB'_r, because the segments A_rB_r and CB'_r lie on the same straight line, and are equal to one another in magnitude and similar in direction.

29. From this property of the straight line r, for the case where all the segments lie upon straight lines which meet in the same point, we obtain a construction for the straight line r in the general case, when the segments lie anywhere upon the plane.

Let C be the point in which A_1B_1 and A_2B_2 intersect. With C as the initial point construct the triangle CDE, whose sides CD, DE are equipollent to the straight lines A_1B_1, A_2B_2; then from what has just been proved for every position of the point O

$$OCE = OA_1B_1 + OA_2B_2.$$

Now let P be the point, in which CE cuts the straight line A_3B_3; with P as initial point construct the triangle PQR, whose sides PQ, QR are equipollent to the segments CE, A_3B_3, then

$$OPR = OCE + OA_3B_3 = OA_1B_1 + OA_2B_2 + OA_3B_3.$$

And so we proceed continually until we ultimately reach a segment AB such that

$$OAB = OA_1B_1 + OA_2B_2 + \ldots + OA_nB_n.$$

This segment AB lies on the required straight line r, and is equipollent to the straight line CN, which joins the extremities of the crooked line $CDE \ldots MN$, whose sides are respectively equipollent to the given segments.

30. As in the general case, when CN is not zero, all the points O, for which the sum

$$OA_1B_1 + OA_2B_2 + \ldots OA_nB_n$$

has the same value, lie upon a fixed straight line (par. 26), so there is only one straight line r, the locus of the points O, for which the above sum is zero. Hence it follows, that whatever be the order, in which we take the given segments in the above construction, we shall always arrive at one and the same straight line r.

CHAPTER II.

GRAPHICAL ADDITION.

31. To geometrically *add* or *combine* a number of segments $1, 2, 3, \dots, n$ given in direction and magnitude, we must construct a polygonal circuit, whose sides, taken in order, are equipollent to the given segments (Fig. 23).

The straight line $s_{1,\dots,n}$ which joins the first and last points of the circuit so constructed, is called the *geometrical sum*

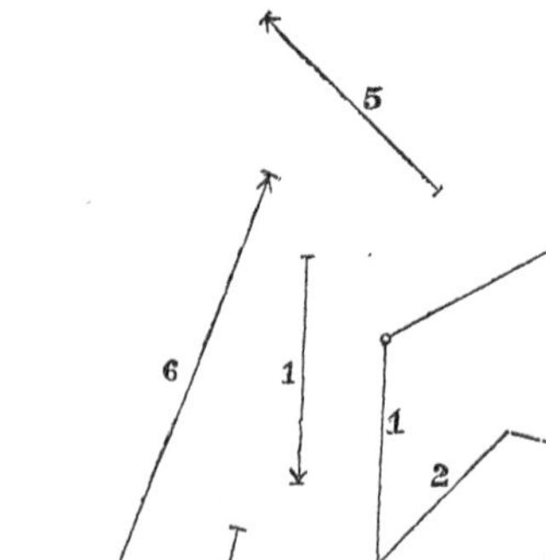

Fig. 23.

Fig. 24.

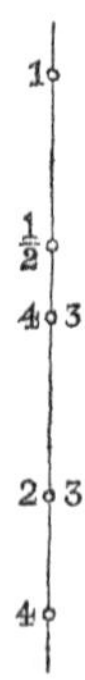

Fig. 25.

or *resultant* of the given segments*; and these are called its *components*. If the given segments are all parallel to one another, the polygonal circuit reduces to a straight line, whose successive segments 01, 12, 23, ... (Fig. 24), or 11, 22,

* CHELINI, *Saggio di Geometria Analitica, trattata con nuovo metodo* (Roma, 1838), p. 35.

33, ... (Fig. 25) are respectively equipollent to the given segments. In this case the resultant of the given segments is identical with their algebraical sum. The two figures show two different methods of denoting a series of segments which follow one another consecutively upon a straight line.

32. From the definition given above, it follows that the resultant $s_{1,\ldots,n}$ of the n given segments $1, 2, 3, \ldots, n$ is identical with the resultant of the two segments $s_{1,\ldots,r}$, $s_{r+1,\ldots,n}$, of which $s_{1,\ldots,r}$ is the resultant of the first r given segments, and $s_{r+1,\ldots,n}$ of the $n-r$ remaining segments. For since the straight lines $s_{1,\ldots,n}$ and $s_{1,\ldots,r}$ start from the same point as the segment 1, and the straight lines $s_{1,\ldots,n}$ and $s_{r+1,\ldots,n}$ end in the same point as the segment n, therefore the straight line $s_{1,\ldots,n}$ begins at the same point with $s_{1,\ldots,r}$ and ends at the same point with $s_{r+1,\ldots,n}$.

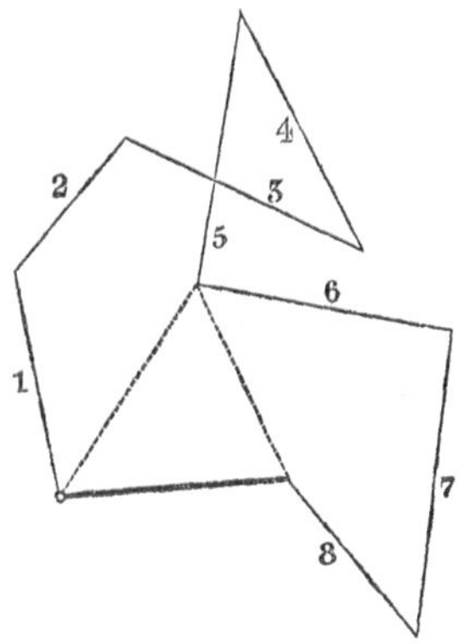

Fig. 26.

Fig. 26 corresponding to $n = 8$, and $r = 5$, shows that the resultant of the segments 1, 2, 3, 4, 5, 6, 7, 8 coincides with the geometrical sum of two components, one of them the resultant of the segments 1, ..., 5, the other the resultant of the segments 6, 7, 8.

From this we infer that, if we divide the given segments [always taken consecutively, i.e. in the given order] into *any* number of groups, and if we sum the segments of each group, the sum of the partial resultants thus obtained will coincide with the resultant of all the given segments.

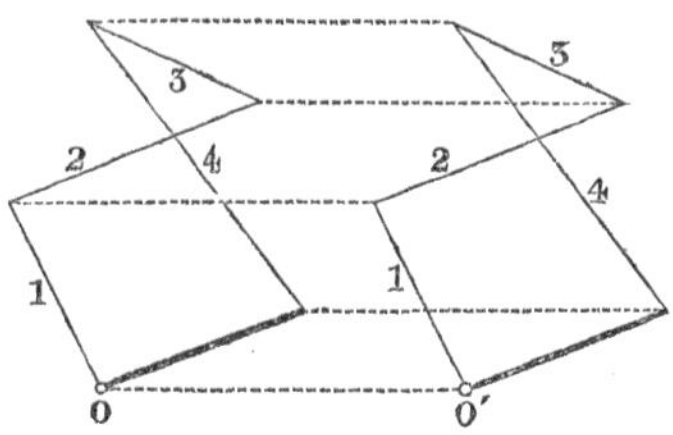

Fig. 27.

33. *The resultant of a number of given segments is independent of the position of the point assumed as the initial point of the circuit.*

In fact the circuits drawn from two different initial points, O and O_1, are equal similar and similarly situated (congruent) figures, and the second may be found by moving the first parallel to itself, so that each of its points describes a straight line equipollent to the straight line OO_1 (Fig. 27).

34. THEOREM. *The resultant $s_{1,\ldots,n}$ of several given segments 1, 2, 3, ..., n is independent of the order in which they are combined.*

Proof. We begin by proving that two consecutive segments, for example 3, and 4 (Fig. 28), can be interchanged.

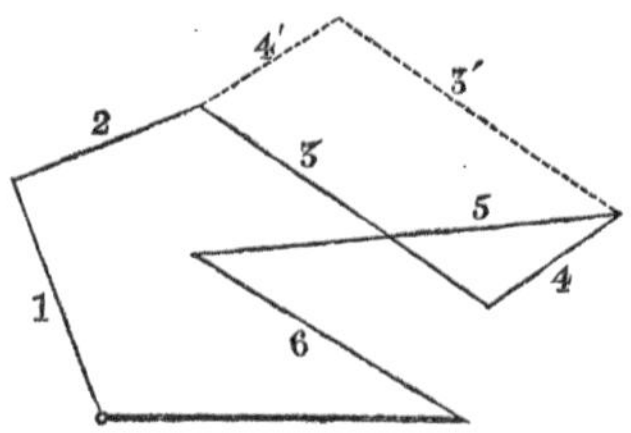

Fig. 28.

In the given order, the resultant of all the segments is also the resultant of the three partial resultants $s_{1,2}$, $s_{3,4}$, $s_{5,\ldots,n}$. In like manner, in the new order, the resultant of all the segments will be the resultant of the partial resultants $s_{1,2}$, $s_{4,3}$, $s_{5,\ldots,n}$. But $s_{3,4}$ and $s_{4,3}$ are the same straight line, namely the diagonal of the parallelogram, which we obtain by drawing first two consecutive segments equipollent to the given ones 3, and 4, and then, starting from the same point, two other consecutive segments equipollent to the same given segments with their order changed 4′3′. Thus the interchange of the segments 3, and 4 has no influence on the required resultant.

If we interchange first 3 and 4, then 3 with 5, and finally 5 and 4, the total effect is the same as if we had interchanged 3 and 5 (Fig. 29). In general we interchange any two non-consecutive segments we please by means of interchanges of consecutive segments. Therefore the resultant of any number of segments is unaltered if we interchange any two segments we please; or the resultant is independent of the order in which the segments are taken to form the figure.

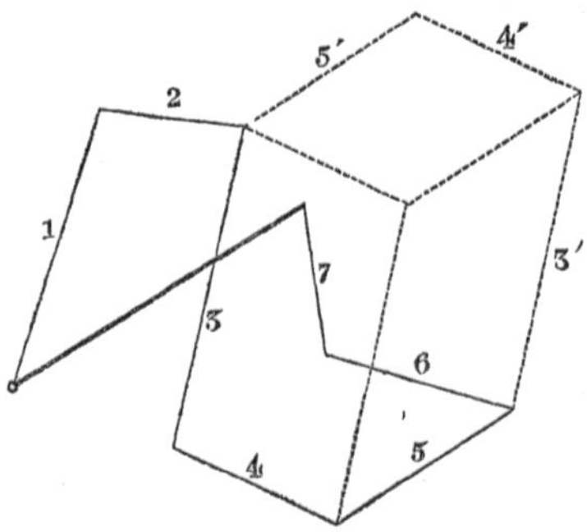

Fig. 29.

Fig. 30 shows several circuits, constructed with the same segments, taken in the different orders 12345, 13254, 15234.

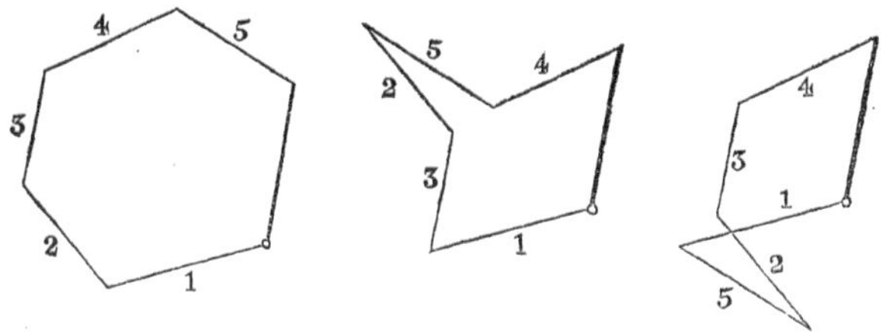

Fig. 30.

35. If a closed circuit can be constructed with the given segments, then from the proposition just proved it follows, that all the circuits obtained by changing the order of the segments have this same property. In this case the resultant of the given segments is zero, or

The resultant of any number of segments vanishes when they are equipollent to the sides of a closed polygon.

The simplest case in which the resultant vanishes is that of only two segments, one of which is equipollent to the other taken in the opposite sense.

36. If, out of some of the segments whose resultant is required, a closed polygon can be formed, then all these may be neglected without affecting the required resultant.

In Fig. 31 the resultant of the segments 1 ... 9 coincides with that of 1, 2, 8, 9, because the resultant of 3, 4, 5, 6, 7 is zero.

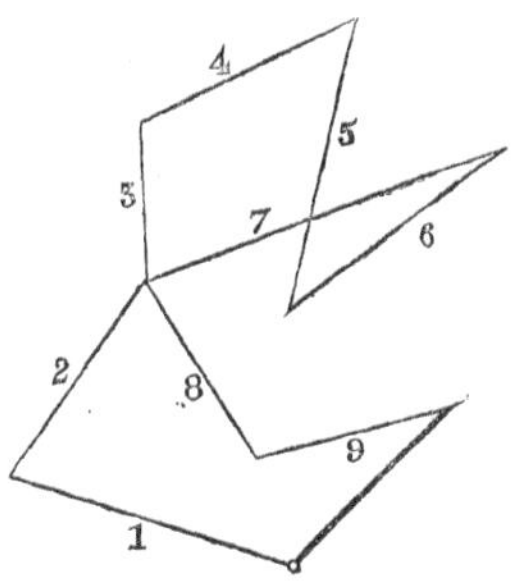

Fig. 31.

If the component segments are increased in any given ratio, then the resultant is increased in the same ratio, without changing its direction.

37. Two series of segments have equal (equipollent) resultants, if, after constructing the corresponding circuits starting from the same point the final points of the two circuits coincide (Fig. 32). If we combine the segments of the one series with those of the other taken in the opposite sense, the total resultant is zero.

38. Two series of segments have equal resultants, but of opposite sense, when, the corresponding polygonal circuits being so constructed that the initial point of the second coincides with the final point of the first, the final point of the second also falls on the initial point of the first. If we combine the two series of segments, their total resultant is zero. Conversely, if the resultant of several segments is zero, and if we split them up into two distinct groups, the resultant of the one group is equal, and of opposite sense, to that of the other group.

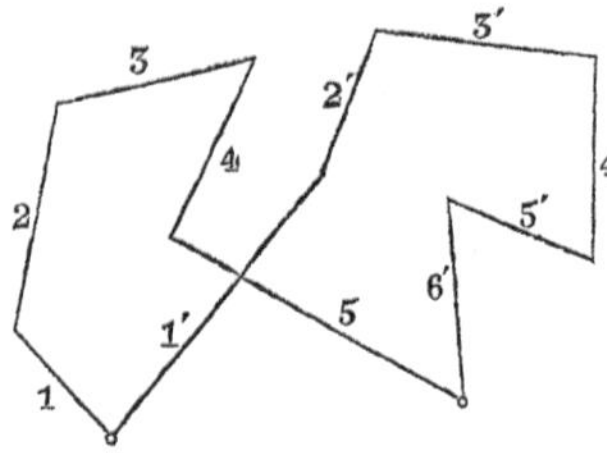

Fig. 32.

39. Subtraction is not a distinct operation from addition. To subtract a segment 1 from a segment 2 is the same as adding to 2 a segment equipollent to the segment 1 taken in the opposite sense.

40. If two series of segments have equal (equipollent) resultants, by adding to, or taking away from, both the same segment, we shall obtain two new series whose resultants will also be equal (equipollent)*.

41. Given a segment AB (Fig. 33), and a straight line r; then if we draw through A and B in any arbitrarily chosen direction two parallel straight lines to meet r in the points A' and B', the points A', B' are called the *projections* of the points A and B, and the segment $A'B'$ the *projection* of the segment AB. The straight lines AA' BB' are called the *projecting rays*.

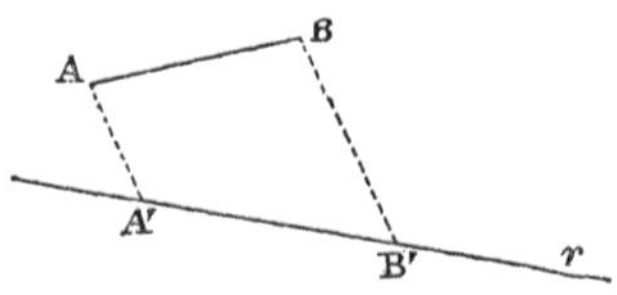

Fig. 33.

The projections of two equipollent segments are themselves equipollent (so long as we neither change the direction of r, nor that of the projecting rays).

* The properties of Art. 32 and Art. 40 can be both deduced without further proof from those of Art. 30.

42. Let $ABC \dots MNA$ be a closed circuit (Fig. 34), and $A', B', C', \dots M', N'$ the projections of its vertices; then since A', B', &c. are points in a straight line, it follows, from (Art. 4), that $A'B' + B'C' + \dots M'N' + N'A' = 0$; i.e. *the sum of the projections of the sides of a closed circuit is zero.*

Let $A_1B_1, A_2B_2, \dots, A_nB_n$ be n segments in a plane, whose resultant is zero, that is to say, n segments which are equal in magnitude and direction to the sides of a closed polygon. Then since the sum of the projections of the sides of a closed polygon is zero, and since the projections of two equipollent segments are equal, therefore the sum of the projections of the given segments will vanish.

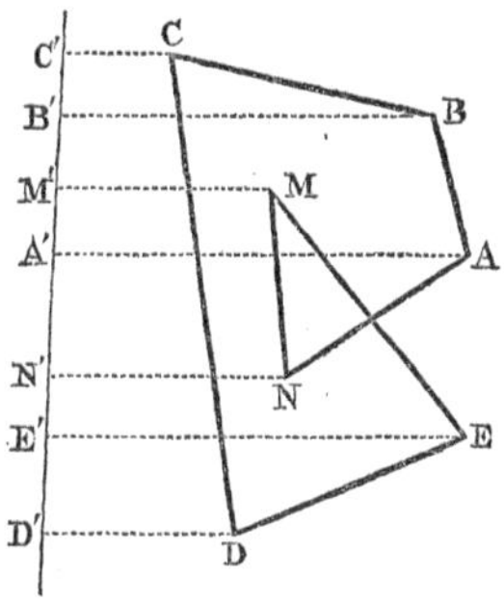

Fig. 34.

A number of given segments together with a segment equal, but of opposite sense, to their resultant, form a system of segments whose resultant is zero. Hence the following proposition:

The projection of the resultant of a number of given segments is equal to the sum of their projections.

From this we at once conclude that:

If two series of segments have equal resultants, the sum of the projections of the segments of the one series is equal to the sum of the projections of the segments of the other.

43. Let $A_1B_1, A_2B_2, \dots, A_nB_n$ be n given segments in a plane, whose resultant is zero (Fig. 35). If we take an arbitrary point O as pole, then we may suppose A_rB_r to be the resultant of the segments A_rO, OB_r; therefore the resultant of the segments $A_1O, OB_1, A_2O, OB_2, \dots, A_nO, OB_n$ will vanish (Art. 38), i.e. the resultant of the segments $OA_1, OA_2, \dots, OA_n$ is equal to that of the segments $OB_1, OB_2, \dots, OB_n$.

Conversely. Given two groups of n points $A_1, A_2, \dots, A_n$, and $B_1, B_2, \dots, B_n$; if the resultant of the straight lines $OA_1, OA_2, \dots, OA_n$, obtained by joining any pole O to the points of the first group, is equal to the resultant of the straight lines $OB_1, OB_2, \dots, OB_n$, got by joining the same pole to the points of the second group: then the resultant of the segments $A_1B_1, A_2B_2, \dots, A_nB_n$, which join the points of the one group to those of the other, is zero. (It is here supposed that the points of the one group can only be properly united to those of the other, when no point is left out, or used more than once.) In fact it follows from the proposition of Art. 37

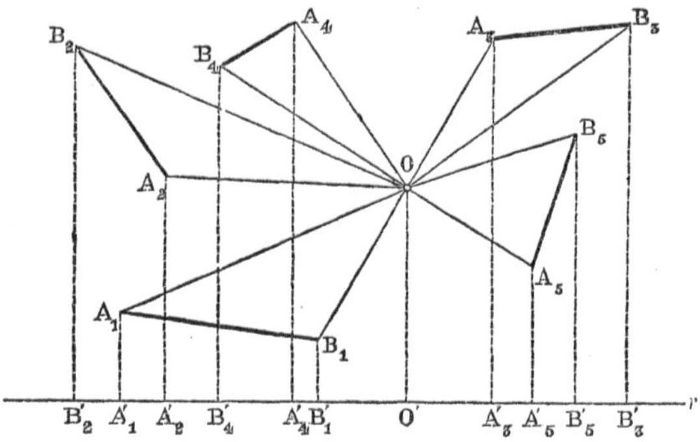

Fig. 35.

that the resultant of the segments A_1O, A_2O, ..., A_nO, OB_1, OB_2, ..., OB_n is equal to zero; but the resultant of A_rO and OB_r is A_rB_r, therefore also the resultant of the segments A_1B_1, A_2B_2, ..., A_nB_n is zero.

44. Hence it follows from the first proposition (Art. 43), that when a new pole O' is assumed, the resultant of the segments $O'A_1$, $O'A_2$, ..., $O'A_n$, is equal to the resultant of the segments $O'B_1$, $O'B_2$, ..., $O'B_n$.

Wherefore *

If, for two groups of n points $A_1, A_2, \ldots, A_n$; $B_1, B_2, \ldots, B_n$ and a fixed pole O, the resultant of the segments OA_1, OA_2, ..., OA_n is equal to the resultant of the segments OB_1, OB_2, ..., OB_n; then the same equality holds for any other pole O'. Moreover the resultant of the n segments, which join the points of the one group with those of the other taken in any arbitrary order, is equal to zero.

45. Retaining the supposition just made as to the two groups of n points, project them into the points A_1', A_2', ..., A_n', B_1', B_2', ... B_n' on a straight line r by means of rays parallel to any arbitrarily chosen direction. Now take the pole O on the straight line r †, then we may suppose the ray OA_r to

* GRASSMANN, *Die Ausdehnungslehre* (Leipzig, 1844), p. 41.

† See Fig. 35, and imagine the straight line r so displaced, that the points O and O' coincide.

be formed by combining the two lines OA'_r, $A'_r A_r$, and so on; the resultant of the segments OA_1', OA_2', ..., OA_n', $A_1'A_1$, $A_2'A_2$, ..., $A_n'A_n$ is therefore equal to the resultant of the segments OB_1', OB_2', OB_3', ... OB_n', $B_1'B_1$, $B_2'B_2$, ..., $B_n'B_n$. But (Art. 41) the resultant or sum of the segments OA_1', OA_2', ... OA_n' is equal to that of the segments OB_1', OB_2', ..., OB_n' since all these segments are the projections of two other series of segments, whose resultants are equal;

Therefore

If for two groups of n points A_1, A_2, &c.; B_1, B_2, &c. and a fixed pole O, the resultant of the segments OA_1, OA_2, &c. is equal to the resultant of the segments OB_1, OB_2, &c., and if we project all the points by means of rays parallel to an arbitrarily chosen direction on to the same straight line, the sum of the projecting rays of the points of the one group is equal to the sum of the projecting rays of the points of the other group.

46. So far, we have been speaking of the resultant of a number of segments, considering only their magnitude, direction, and sense, but not their absolute position. We shall now give a more general definition, which includes the one previously given (31), and takes account of all the elements of the resultant straight line of a number of given segments.

If n segments A_1B_1, A_2B_2, ..., A_nB_n, are given (in sense, position, and magnitude) their *resultant* will mean a segment AB of such magnitude, position, and sense, that, for any pole O, the area of OAB is equal to the sum of the areas $OA_1B_1 + OA_2B_2 + \dots + OA_nB_n$ (26, 30).

47. For shortness we shall call the triangle OAB, *the triangle which joins the segment AB to O.* The sense AB of this segment shows the way in which the circuit OAB is traced out, and therefore shows the sense of the area OAB.

This being premised, our definition may be expressed as follows. *By the resultant of a number of given segments, we mean a segment such that the area of the triangle which joins it to an arbitrary pole O, is equal to the sum of the areas of the triangles which join the given segments to the same pole.*

Since the area of the triangle OAB does not change, if we displace the segment AB along the straight line on which it lies, therefore the resultant of a number of segments will not change, if we displace each of them in an arbitrary manner along the straight lines on which they respectively lie.

48. We know already from Art. 26 that if we construct a polygonal circuit $CDE \ldots MN$, the sides of which are respectively equipollent to the given straight lines $A_1 B_1, \ldots, A_n B_n$, then its closing side NC is equipollent to the resultant AB. If the circuit is closed, i. e. if N coincides with C, but if the sum of the areas $OA_1 B_1 + OA_2 B_2 + \ldots OA_n B_n$ is not zero, then the magnitude of the required resultant is zero and it is situated at an infinite distance. If the circuit is closed, and the above sum also zero, then the magnitude of the resultant is still zero, and its position is indeterminate. In this case therefore it may be asserted that the given series of segments has no resultant.

49. But if C does not coincide with N, then the problem is uniquely solved by a segment AB of finite magnitude, situated at a finite distance. As we already know its magnitude, its direction, and its sense, it will be sufficient, in order to completely determine its position, to find one point in the straight line of which it forms a part. For this purpose we may use either the construction in Art. 29, or else the much simpler one following (Fig. 36).

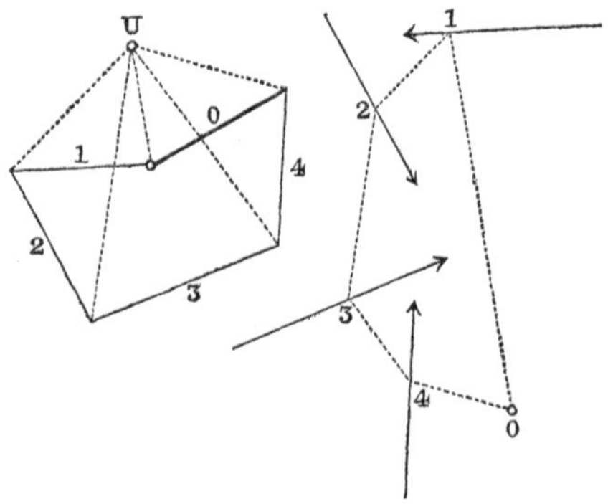

Fig. 36.

We begin by constructing a polygonal circuit, with its sides, which we shall now denote by 1, 2, ..., n, respectively, equipollent to the given segments; then their resultant is equipollent to the segment 0, which closes the circuit, taken in the opposite sense, i.e. it is equal, but of opposite sense, to the segment which joins the final point of the side n to the initial point of the side 1. We now choose at pleasure a pole U, and draw from it the rays UV_{01}, UV_{12}, ..., UV_{n0}* to the vertices of the circuit; where $V_{i\,.\,i+1}$ means the vertex which is the final point of the side i (equipollent to A_iB_i), and the initial point of the side $i+1$ (equipollent to $A_{i+1}B_{i+1}$).

We next construct a second polygonal circuit with its vertices 1, 2, ..., n lying respectively on the lines to which the segments A_1B_1, A_2B_2, ..., A_nB_n belong, and with its sides 01, 12, ..., $n0$ respectively parallel to the rays UV_{01}, UV_{12}, ..., UV_{n0}. The extreme sides 01, $n0$ of this polygon will, if sufficiently produced, meet in a point 0 which lies on the required line of the resultant †.

Proof. We suppose the segment A_1B_1 to be resolved into 2 others, situated in the sides 01, 12 of the second polygon, and equipollent to the rays $V_{01}U$, UV_{12} of the first; in like manner we suppose the segment A_2B_2 resolved into two others, situated in the sides 12, 23 of the second polygon, and equipollent to the rays $V_{12}U$, UV_{23} of the first; and so on, till finally A_nB_n is resolved into two segments situated on the sides $n-1\cdot n$, $n0$, and equipollent to the rays $V_{n-1\,.\,n}U$, UV_{n0}.

If we take any pole O, then the area of the triangle, which joins it to one of the given segments, is equal to the sum of the two triangles which join its two component segments to the same pole; and consequently the resultant of the n given segments A_1B_1, A_2B_2, ..., A_nB_n coincides with the resultant of the 2 n component segments into which the given ones have been resolved. Now the first of these 2 n segments is situated on 01, and equipollent to $V_{01}U$, and the last is situated on $n0$, and equipollent to UV_{n0}, whilst all the rest, 2 $(n-1)$ in number, are equal to one another in pairs, are of opposite sense, and are situated on the same side of the second polygon. For example, the second and third component segments lie on the side 12, and are respectively equipollent to UV_{12} and $V_{12}U$.

The areas of the two triangles, which join these pairs of segments to O, are equal to one another but of opposite sense; the resultant of the given segments is therefore no other

* In Fig. 36 all the letters V, A, B are left out, and $n = 4$.

† Culmann, l. c., Nos. 41 & 42.

than the resultant of the first and last component segments, of which the first is situated in 01 and equipollent to $V_{01}U$, and the other is situated in $n0$ and equipollent to UV_{n0}. But the resultant of two segments passes through the common point of (Art. 28) the straight lines to which they belong, therefore the required resultant passes through the common point of the two extreme sides 01, $n0$ of the second polygon.

50. If the pole U were taken in a straight line with the two extreme points V_{01}, V_{n0} of the first polygon, then the two extreme rays UV_{01}, UV_{n0} would coincide, and therefore the two extreme sides 01, $n0$ of the second polygon would be parallel. In this case therefore the construction would not give a point at a finite distance in the required resultant. But this inconvenience could at once be remedied by choosing a new pole U' not lying in the straight line V_{01}, V_{n0} and then proceeding as above.

51. Even if that is not so, it may happen (Fig. 37) that the points V_{n0} and V_{01} coincide, and then, wherever U may be,

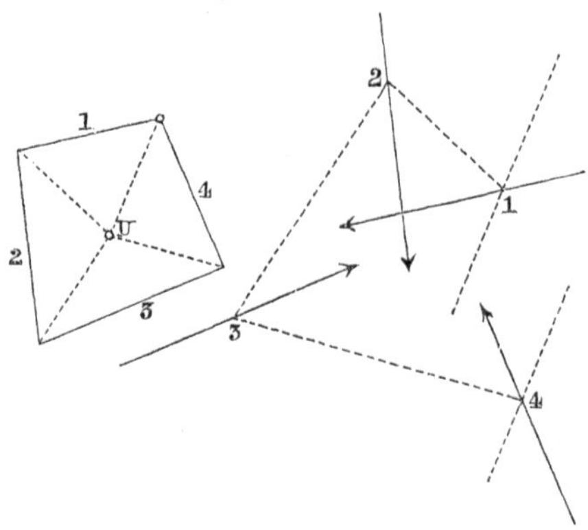

Fig. 37.

the extreme rays coincide, and therefore the sides 01, $n0$ are either parallel or coincident. If they are parallel, the sum $OA_1B_1 + OA_2B_2 +$ &c. is equal to the sum of the two triangles whose common vertex is O, and whose bases are equipollent to the equal and opposite rays $V_{01}U$, UV_{01} and lie in the sides 01, $n0$, or is equal to the half of the parallelogram (Art. 15), of which those bases are the opposite sides.

In this case the resultant is zero and situated at an infinite distance; and the sum OA_1B_1+&c. has a constant value not zero, wherever (at a finite distance) the pole O may lie.

52. If, on the contrary, the sides (Fig. 38) 01, $n0$ coincide, i.e. if the opposite sides of the parallelogram coincide, then the sum $OA_1B_1+OA_2B_2+$&c. vanishes for every position of the pole O. In this case any one segment taken in the reverse sense is the resultant of the remaining $(n-1)$ segments.

53. If we take the given segments $A_1B_1, A_2B_2, \ldots$ all parallel to one another, then the first polygon $V_{01}\ V_{12}\ V_{23} \ldots V_{n0}$ (Fig. 39) reduces to a straight line, but the construction of the

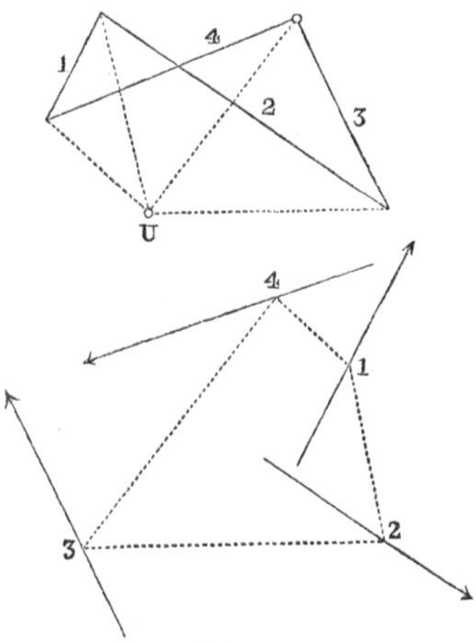

Fig. 38.

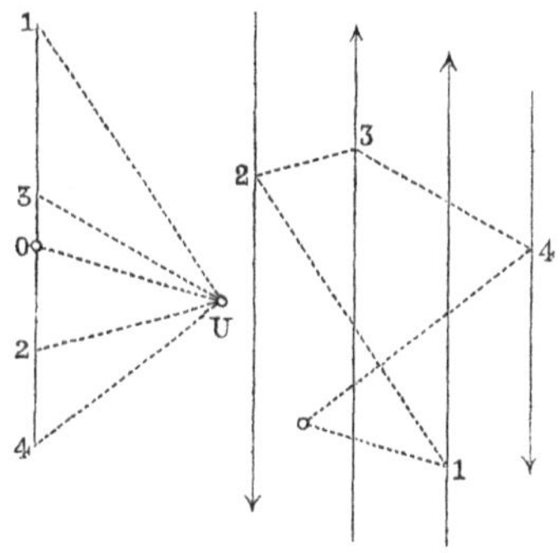

Fig. 39.

second polygon is just the same as in the general case. The resultant is parallel to the components.

54. If there are only two segments A_1B_1, A_2B_2, the construction may be simplified as follows (Figs. 40, 41). In the unlimited straight line $A_1B_1 \ldots$ take a segment CD equipollent to A_2B_2, and in the unlimited straight line $A_2B_2 \ldots$ a segment $C'D'$ equipollent to A_1B_1. Then the common point O of the straight lines CD', and $C'D$ lies on the required resultant. For if we draw $D'E$ parallel to $C'D$ and join O to E, then C, D, E represent the vertices V_{01}, V_{12}, V_{20} of the first polygon, and O takes the place of the point U; the points D', E are the vertices 1, 2 of the second polygon, which is here represented by the triangle $OD'E$, and O represents also the point of intersection of the extreme sides of this second polygon.

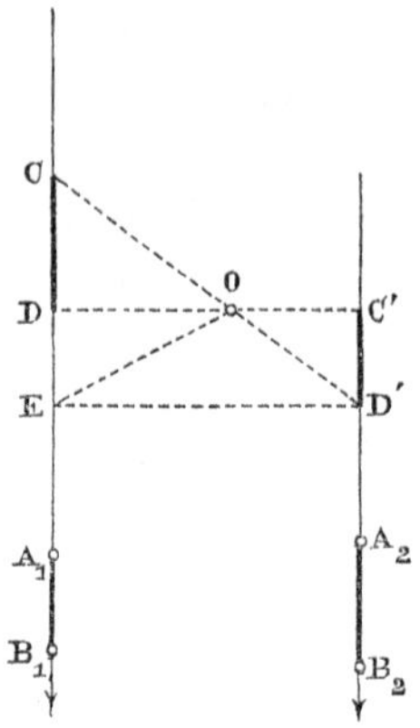

Fig. 40.

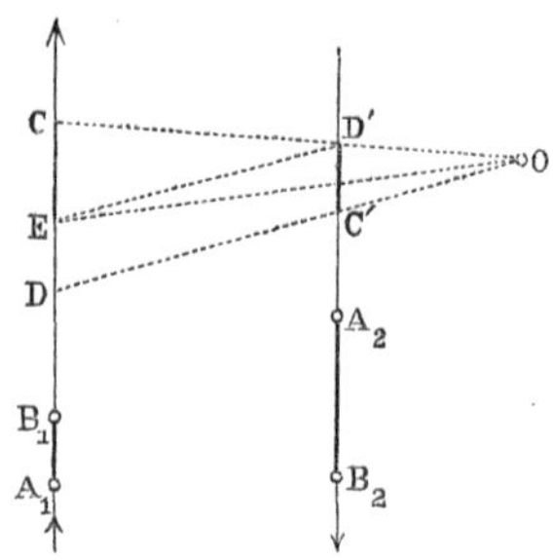

Fig. 41.

From the similar triangles OCD, $OD'C'$ we have

$$OC' : OD = C'D' : DC$$
$$= A_1B_1 : B_2A_2 ;$$

that is to say:

The ratio of the distances of the resultant of two parallel segments from these segments is the negative reciprocal of the ratio of the component segments.

CHAPTER III.

GRAPHICAL MULTIPLICATION.

55. To *multiply* a straight line a by the ratio of two other straight lines $b:c$, we must find a fourth straight line x such that the geometrical proportion holds:

$$c:b = a:x.$$

For this purpose it is sufficient to construct two similar triangles OLM and $O'PQ$ with the following properties.

In the first there are two lines (two sides, or base and altitude, and so on) equal or proportional to c, b; and in the second the line homologous to c is a; then x is the line in the second triangle homologous to b; or else

In the first there are two lines proportional or equal to c and a; and in the second the line homologous to c is b; then x is the line of the second triangle homologous to a.

56. The relative position of the two triangles is purely a matter of choice; and the particular choice made gives rise to different constructions. The choice will be chiefly determined by the position occupied by the given segments a, b, c, or of that which we wish x to occupy.

(a) In (Fig. 42), for example, the two triangles have the angle O in common and the sides opposite to it parallel. If in them

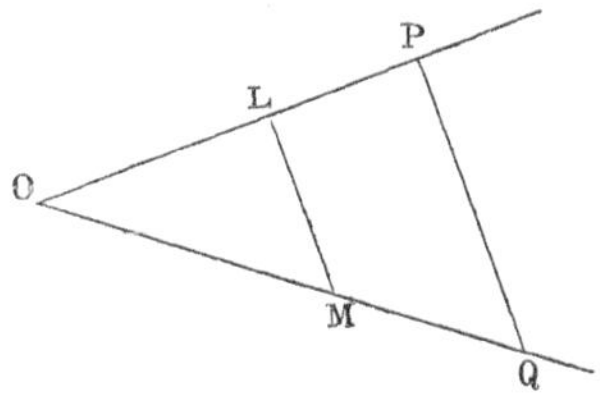

Fig. 42.

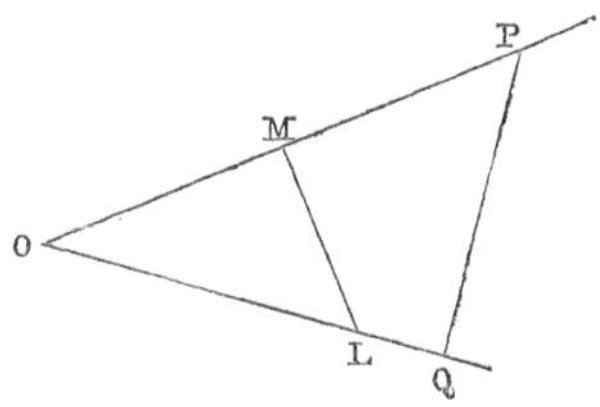

Fig. 43.

we take OP, OM, OL to represent the segments a, b, c, then $OQ = x$. But if $OL = c$, $OP = a$, $LM = b$, then $PQ = x$.

(*b*) In (Fig. 43), on the contrary, the sides opposite to the common angle O are *antiparallel*, i.e. the angles OML and OQP are equal (and therefore also the angles OLM and OPQ).

(*c*) (Fig. 44). We may take c and a to be the altitudes of the two triangles; and then, on the supposition that b is a side OM or LM of the first triangle, OQ or PQ will be equal to x.

(*d*) Or again, let c and a be represented by OL, OP, or by OM, OQ, and let b be the altitude of the triangle OLM, then x is the altitude of the triangle OPQ.

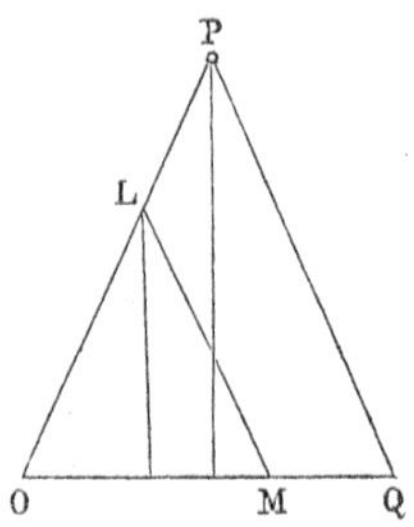

Fig. 44.

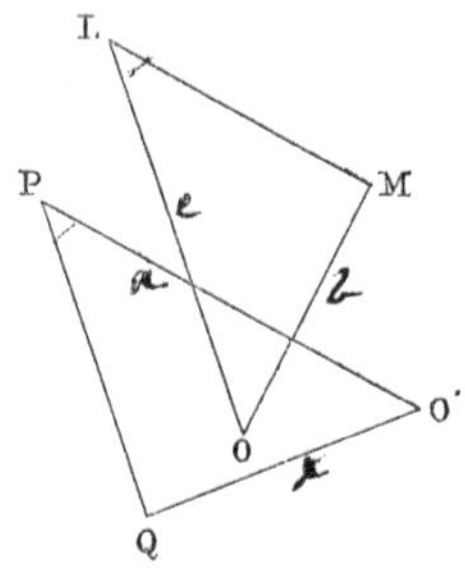

Fig. 45.

(*e*) If (Fig. 45) the lines $OM = b$ and $O'P = a$ are drawn perpendicular to one another, supposing that $c > b$, we may proceed as follows. Construct the triangle OLM, so that the side LM is parallel to $O'P$, whilst the hypothenuse $OL = c$. Then if we draw PQ parallel to OL, and $O'Q$ perpendicular to PQ, the right-angled triangles OLM, $O'PQ$ are similar because of the equal angles L and P; and therefore $O'Q = x$.

The straight line $O'Q$, the orthogonal projection of $O'P$ upon a straight line at right angles to OL, is called the *Antiprojection* of $O'P$ on OL. If therefore a and b are perpendicular to one another, then x is the antiprojection of a upon c.

57. To *divide* a straight line a by the ratio of two other straight lines $b : c$, is just the same as to multiply a by the ratio $c : b$.

The division of a straight line a into n equal parts is the same thing as multiplying a by the ratio $c : b$, where c is an arbitrary segment, and b is equal to c repeated n times.

If a straight line b is to be divided into parts, which are proportional to the given segments a_1, a_2, $a_3, \ldots, a_n$, lying in the same straight line, then we have only to multiply these segments by the ratio $b : c$, where $c = a_1 + a_2 + \ldots + a_n$ (Art. 59).

58. If we draw from a centre or pole O radii vectores, each consecutive pair of which contains the constant angle ω, their lengths forming the Arithmetical Progression,

$$a,\ a+b,\ a+2b,\ \&c.,\ \&c.;$$

then their extremities M, M_1, M_2, &c. are points on a curve, called the *Spiral* of *Archimedes*; which is the name given to the curve described by a point M which moves uniformly along the radius OM whilst the radius itself rotates about O with constant velocity, in such a manner that M describes the rectilinear segment b in the same time that the radius OM describes the angle ω.

If we take the angle ω small enough, we shall obtain points sufficiently close together to be able to draw the curve with sufficient accuracy for all practical purposes.

After we have drawn the *Spiral* of *Archimedes*, we are able to reduce the problem of dividing an angle to that of the division of a straight line. For if two radii vectores are drawn, which enclose the angle we wish to divide into n parts proportional to n given straight lines, we need only divide the difference of the radii vectores into n parts proportional to the same magnitudes; and then the distances of the point O from the $n-1$ points of division will be the lengths of the $n-1$ radii vectores to be inserted between the two given ones, in order to obtain the division of the angle. Fig. 46 shows the division of the angle MOM_5 into five equal parts*.

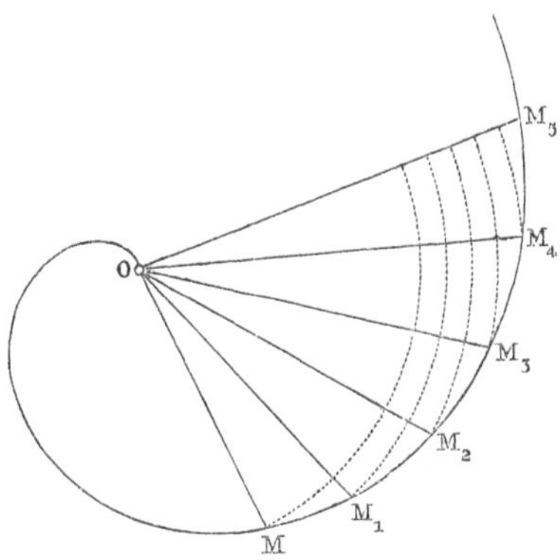

Fig. 46.

59. If several segments AB, AC, …, BC, … of a straight line u have to be multiplied by a constant ratio $b : c$, the problem

* Pappus, *Collectiones Mathematicae*, Lib. iv. Prop. xx, xxxv.

resolves itself into finding a series of points A', B', C', &c. on another straight line u'; such that the equations

$$\frac{A'B'}{AB} = \frac{A'C'}{AC} = \frac{B'C'}{BC} = \ldots = \frac{b}{c} \text{ hold.}$$

The straight lines u, u' are called *similar point-rows*, and the points A and A', B and B',..., and also the segments AB and $A'B'$..., are said to be *corresponding*.

60. If the straight lines u, u' are parallel (Fig. 47), then all the joining lines AA', BB', CC', &c., pass through a fixed point O (the centre of projection). If, for example, we make $AP = c$, $A'P' = b$, then AA' and PP' give by their intersection the point O, and every radius vector drawn through the point O cuts u and u' in two corresponding points.

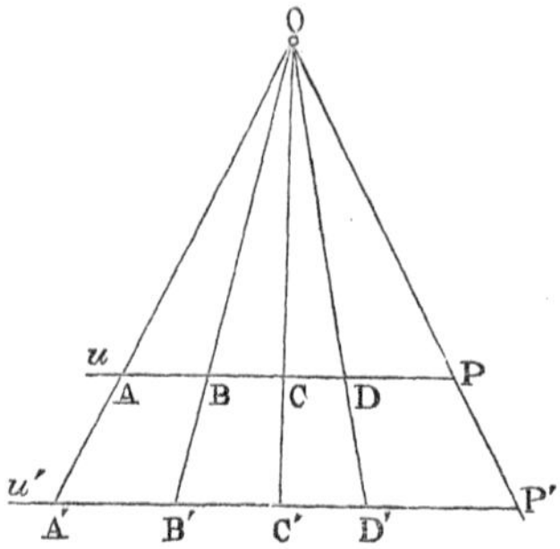

Fig. 47.

61. If u and u' are not parallel (Fig. 48), and if their common point represents two coincident corresponding points A and A', then the straight lines BB', CC', &c. are all parallel to one another. The common direction of these parallel lines may be

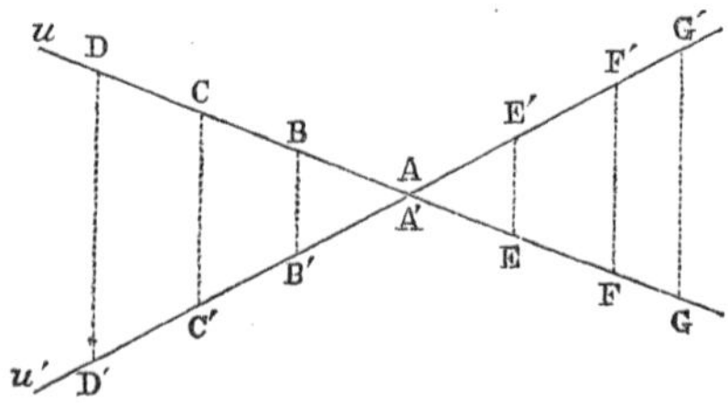

Fig. 48.

found, for example, by taking $AB = c$, $A'B' = b$; then every straight line parallel to BB' cuts u and u' in two corresponding points.

62. If, finally (Fig. 49), u and u' are not parallel, and their common point represents two non-corresponding points P, Q', then all the straight lines AA', BB', CC', ... are tangents to the same parabola. If, for example, we take $PQ = c$, $P'Q' = b$, then the parabola is determined from the fact that, it must touch u in Q, and u' in P'. Every tangent of this parabola cuts u' and u in two corresponding points.

In order to obtain pairs of corresponding points, such as A and A', B and B', &c., we need only draw from the different

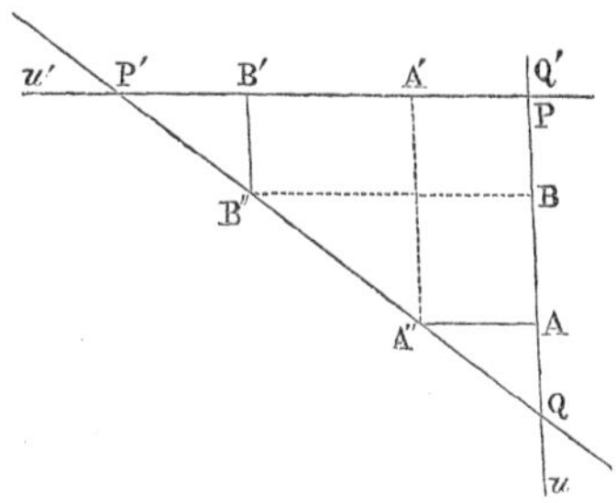

Fig. 49.

points A'', B'', &c. of the straight line $P'Q$ the straight lines $A''A$, $B''B$, &c. parallel to u', and the straight lines $A''A'$, $B''B'$, &c. parallel to u. For then clearly we have

$$\frac{A'B'}{A''B''} = \frac{P'Q'}{P'Q}, \quad \frac{AB}{A''B''} = \frac{PQ}{P'Q};$$

and therefore

$$\frac{A'B'}{AB} = \frac{P'Q'}{PQ} = \frac{b}{c}$$

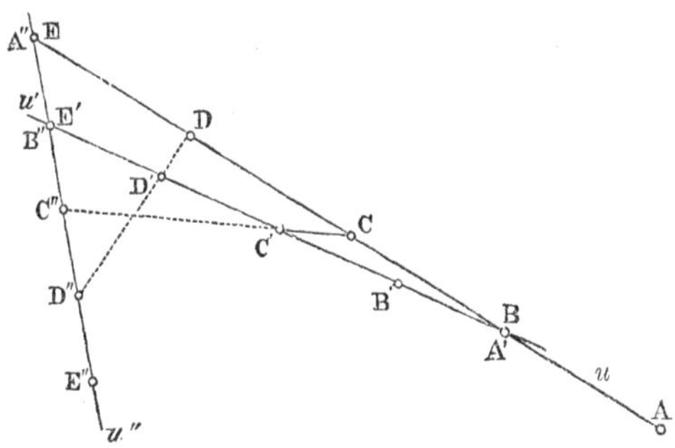

Fig. 50.

If we wish to avoid drawing parallels* it is sufficient to consider two tangents (Fig. 50) of the parabola as given, i. e. two straight lines u, u'', in which two similiar point-rows (they

* Cousinery, *Le calcul par le trait* (Paris, 1840), p. 20. For another method of solving this problem see Sacheri, *Sul tracciamento delle punteggiate projettive simili* (Atti dell' Accademia di Torino, Novembre, 1873).

may be equal) $ABCDE, \ldots, A''B''C''$, &c. are so situated that the common point of the two straight lines represents two non-corresponding points E, A'', and that the segment AE of u [which is contained between the parabola and u''] is equal to the denominator c of the given ratio. If we want now to multiply the segments of u by the ratio $b : c$, we must place a line $A'E'$ of length b between u and u'' in such a manner that it joins two corresponding points BB''. Then the straight lines CC'', DD'', &c. which join corresponding points of u and u'', determine upon $A'E'$ the required segments

$$B'C' : C'D' : D'E' : A'E' =$$
$$BC : CD : DE : AE.$$

If, for example, it were required to divide a given length BE into n equal parts, we should draw through B the straight line u', and lay off on it $n+1$ equal segments,

$$A'B' = B'C' = C'D', \text{ \&c.};$$

then having joined E to E' we should, in like manner, lay off upon the joining line u'', $n+1$ segments each equal to EE', or $A''B'' = B''C'' = C''D'' =$ &c., &c. The $n+1$ straight lines $C'C''$, $D'D''$, &c., &c. will meet BE in the required division-points C, D, &c.

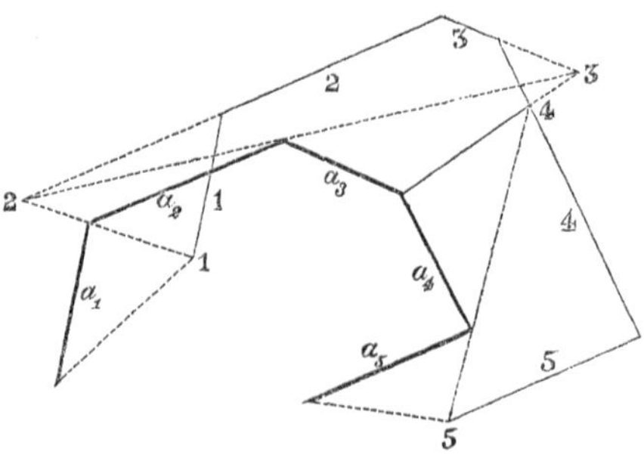

Fig. 51.

63. Let (Fig. 51) $a_1, a_2, \ldots, a_n$ be n segments given in magnitude, direction, and sense, which have to be respectively multiplied by the ratios

$$\frac{b_1}{c_1}, \frac{b_2}{c_2}, \text{ \&c., } \frac{b_n}{c_n}.$$

We construct a polygonal circuit P_a, whose sides are respectively equipollent to the given segments a_1, a_2, &c., and call its successive vertices 1, 12, 23, ... $n-1.n$, n, beginning at the initial point of the first side a_1 and ending with the final point of the last side a_n.

Then we construct two other circuits P_c and P_{ac}; of which the first is formed by the n straight lines 1, 2, ..., n respectively parallel to the sides of P_a, and at the respective distances c_1, $c_2, \ldots, c_n$ from them, each measured in a constant direction which may be fixed arbitrarily, provided that c_r be not parallel to a_r *; and the second P_{ac} must have its vertices 1, 2, ..., n respectively upon the sides of P_c, and its sides 1, 12, 23, ..., $n-1.n$, n† must respectively pass through the similarly named vertices of P_a. The combination of these three circuits is called 'the First Figure.'

Now construct a 'Second Figure,' which similarly consists of three circuits P_x, P_b, P_{xb}, having the following properties (Fig. 51 *a*):

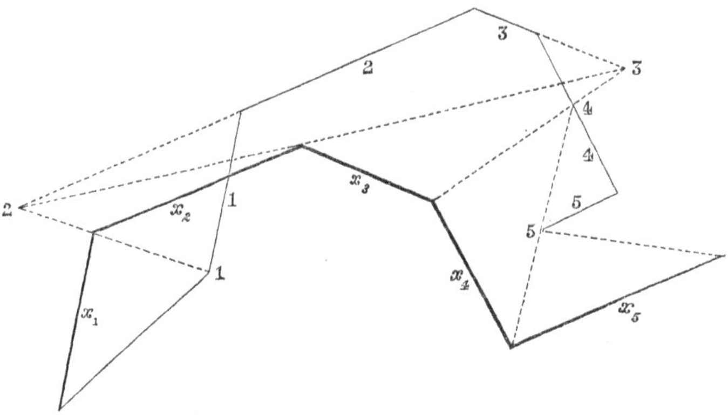

Fig. 51 *a*.

1. The sides of P_x are respectively parallel to the sides of P_a; the sides of P_b parallel to those of P_c (and therefore to those of P_a and P_x); the sides of P_{xb} to those of P_{ac}.

2. The distances of the sides 1, 2, ..., n of P_b from the similarly named ones of P_x are $b_1, b_2, \ldots, b_n$ measured parallel to the distances $c_1, c_2, \ldots, c_n$.

3. Each of the vertices 1, 2, ..., n of P_{xb} lies on the

* According as c_r is positive or negative, we draw the straight line r to the right or left of a person who travels along a_r in the sense belonging to that segment.

† The side 1 is that which goes through the vertex 1; the side 12 joins the vertices 1, 2; ...; the side n passes through the vertex n. In order to construct this polygon, we can take the side 1 at pleasure, provided it passes through the vertex 1 of P_a.

similarly named side of P_b, and each of its sides $1, 12, 23, \ldots, n-1\,n, n$ must pass through the similarly named vertex of P_x.

In order to construct the 'Second Figure' we may, for example, proceed thus. The vertex 1 of P_x is taken arbitrarily, and through it two straight lines are drawn, respectively parallel to the side a_1 of P_a, and the side 1 of P_{ac}. These determine the positions of the side 1 of P_x and the side 1 of P_{xb}. If now, at a distance b_1 from the side 1 of P_x, a straight line is drawn parallel to this side, then this line will be the first side of P_b, and the point where it meets the side 1 of P_{xb} will be the vertex 1 of P_{xb}.

From this point draw (parallel to the side 12 of P_{ac}) the side 12 of P_{xb}, then the intersection of this with the side 1 of P_x, will be the vertex 12 of P_x. From this point we draw the side 2 of the polygon P_x in the direction of the segment a_2, and afterwards the side 2 of P_b in the same direction, but at a distance b_2 from it, then the intersection of this with the side 12 of P_{xb}, gives the vertex 2 of P_{xb}, and so on.

The polygon P_x, whose sides we shall call $x_1, x_2, \ldots, x_n$, gives the result of the required multiplication. For the triangle, which has x_r for its base, and for its opposite angle the vertex r of P_{xb}, is, on account of the parallelism of the sides, similar to the triangle of the 'First Figure,' which has a_r for base, and the vertex r of P_{ac} for the opposite angle. The dimensions of these triangles in the chosen directions are b_r, c_r, and therefore

$$x_r : a_r = b_r : c_r \text{ or } x_r = a_r \times \frac{b_r}{c_r}.$$

Q. E. D.

64. With regard to the sense of the segment x_r we remark, that the two triangles are similarly situated when c_r and b_r have the same sense, i.e. when the vertices r lie both to the right or both to the left of the bases (a_r or x_r) respectively opposite to them; if, on the contrary, c_r and b_r are of opposite sense, then the two triangles have opposite positions. For this reason in the first case the segments a_r and x_r are of the same sense, in the second case of opposite sense. Hence it follows, that the segments x are placed consecutively taking account of their sense, i.e. in the way which is required by Geo-

metrical addition. Wherefore their resultant, i.e. the resultant of the segments $a_r \cdot \frac{b_r}{c_r}$, will be, in magnitude, sense, and direction, the straight line which closes the polygonal contour P_x (i.e. the straight line which joins the initial point of x_1 to the final point of x_n).

65. Special cases.

Let all the segments a become parallel; then each of the two circuits P_a and P_x reduces to a rectilinear point-row (Fig. 52)

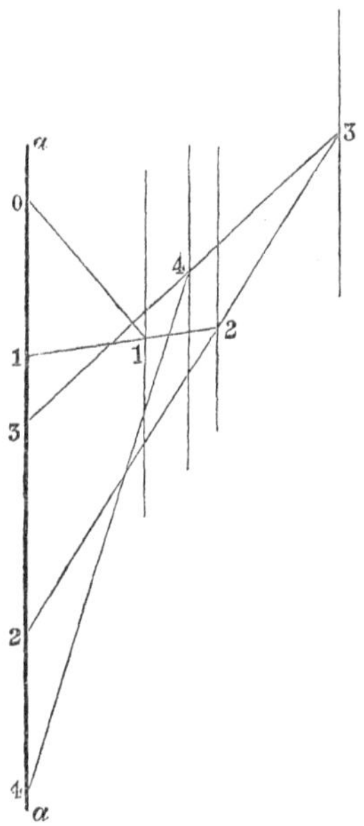

Fig. 52.

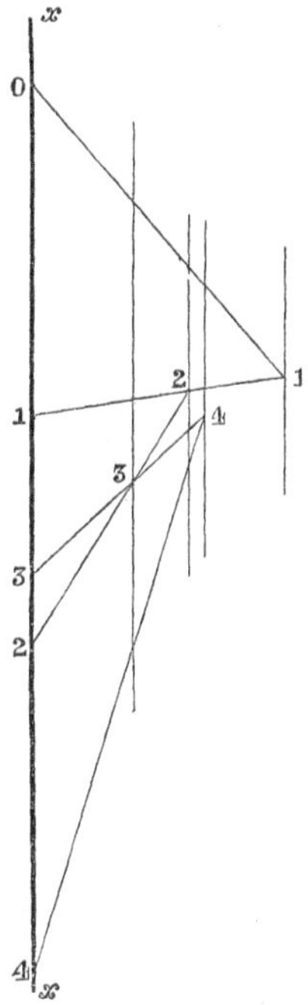

Fig. 52 a.

and each of the circuits P_c and P_b becomes a pencil of parallel rays. That is to say, the construction reduces to the following.

We set off the consecutive segments $01 = a_1$, $12 = a_2$, $23 = a_3$, &c., along a straight line a; parallel to this line and at distances $c_1, c_2, c_3, \ldots, c_n$ (measured in some constant direction, different from the direction of the a's, but otherwise arbitrary) we draw as many straight lines 1, 2, ..., n, which we may consider as rays of a pencil whose centre lies at infinity;

then draw a polygonal circuit, with its vertices $1, 2, \ldots, n$ on the similarly named parallel rays, and with its sides $01, 12, 23, \ldots, n-1\ n, n$ passing through the corresponding points $0, 1, 2, \ldots, n$ of the point-row a [i.e. through the extremities of the segments $a_1, a_2, \ldots, a_n$].

Now construct the second figure, by drawing, first, a pencil of rays $1, 2, \ldots, n$, parallel to a, and at distances $b_1, b_2, \ldots, b_n$ respectively from a straight line x (also parallel to a); and then a circuit whose sides are respectively parallel to the sides of the first polygon, and whose vertices fall on the rays of the second pencil. The segments 01, 12, 23, &c. of x, which are enclosed between the successive sides of this new polygon, will be

$$x_1 = a_1 \cdot \frac{b_1}{c_1}, \quad x_2 = a_2 \cdot \frac{b_2}{c_2}, \quad x_3 = a_3 \cdot \frac{b_3}{c_3}, \text{ \&c., respectively,}$$

and the segment, that lies between the side $r-1 \cdot r$ and the side $s \cdot s+1$, is equal to

$$\sum_{i=r}^{i=s} x_i = \sum_{i=r}^{i=s} a_i \cdot \frac{b_i}{c_i} *.$$

In the case just considered it is an immediate deduction from the remarks made about the sense of the segment x_r, that two segments x_r, x_s have the same or opposite sense, according as amongst the three pairs $a_r\ a_s$, $b_r\ b_s$, $c_r\ c_s$, an even number (none or two) or an odd number (one or three) are formed by segments of opposite sense. This agrees with the rule of signs in algebraic multiplication.

66. If all the c's become equal, in addition to all the a's being parallel, then the first pencil of rays reduces to a single straight line, and therefore all the vertices of the first polygon coincide in a single point of this straight line; i.e. the first polygon degenerates into a pencil of rays proceeding from a point O situated at a distance c from the straight line a.

In this case the problem may be stated thus. To reduce the given products $a_1 . b_1, a_2 . b_2, \ldots, a_n . b_n$ to a common base c, by determining the segments $x_1, x_2, \ldots, x_n$ proportional to them.

The solution is as follows (Fig. 53). Draw the rectilinear point-row a, whose consecutive segments are $01 = a_1$, $12 = a_2$,

* JAEGER, *Das Graphische Rechnen* (Speyer, 1867), p. 15.

&c., &c., and join each of the points 0, 1, 2, ..., $n-1$, n of this point-row to a point O taken at a distance c from a; the distance being measured perpendicularly, or obliquely

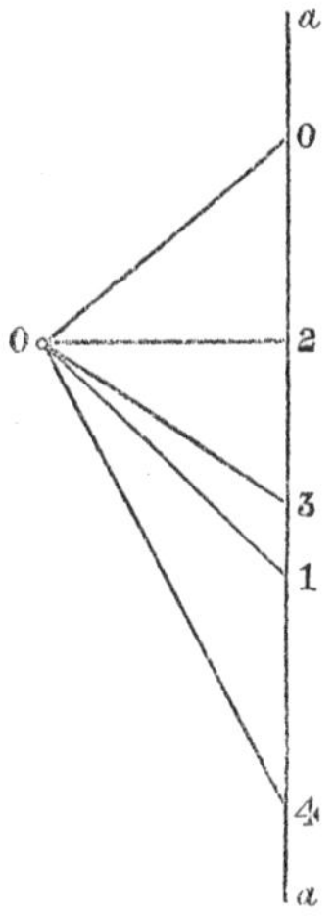

Fig. 53.

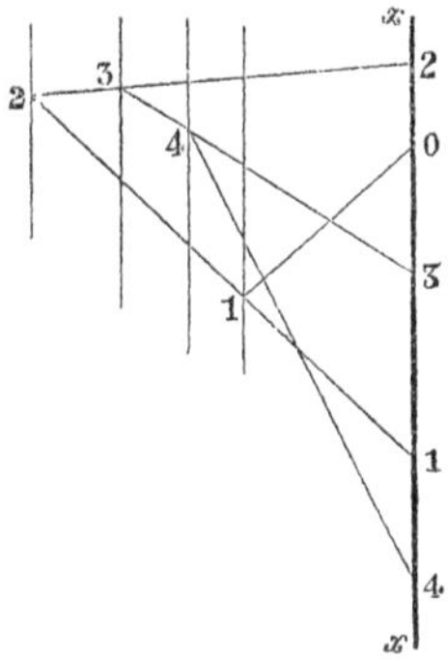

Fig. 53 *a*.

at pleasure. Then construct a pencil of rays 1, 2, ..., n, parallel to a, and at distances b_1, b_2,..., b_n, respectively, measured in the direction of c from a given line x, also parallel to a. Finally, draw a polygon whose vertices fall respectively on the above-mentioned parallel rays 1, 2, ..., n, and whose sides 01, 12, 23, ..., $n-1.n$, n are respectively parallel to the rays $O0$, $O1$, $O2$, ..., $On-1$, On of the pencil O. The segments 01, 12, 23, ..., which the sides of this polygon intercept upon the straight line x, will be the required segments x_1, x_2, ..., x_n*.

67. If instead of all the c's, all the b's are equal, and all the a's still parallel, then the problem may be stated thus:

Given the ratios

$$\frac{a_1}{c_1}, \frac{a_2}{c_2}, \ldots, \frac{a_n}{c_n}$$

to determine segments x_1, x_2, x_3, ..., x_n, proportional to them, so that the product of the multiplication of any x by its corresponding ratio $\frac{c}{a}$ shall be the constant segment b.

* CULMANN, l. c., No. 2.

After we have constructed (Fig. 54) the point-row a, with the segments $01 = a_1$, $12 = a_2$, &c., and the pencil of rays $1, 2, 3, \ldots, n$ parallel to the straight line a, their respective distances from it being c_1, c_2, c_3, &c. (all measured in a constant direction), we draw a polygonal circuit, whose vertices $1, 2, \ldots, n$ fall on these rays respectively, and whose sides $1, 12, 23, \ldots, n-1\ n, n$ pass through the similarly named points of the point-row a.

We then construct a second pencil of rays, diverging from a point O, and respectively parallel to the sides of the poly-

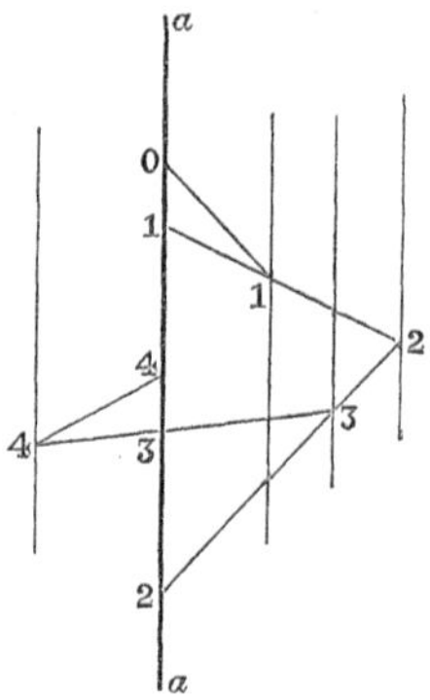

Fig. 54.

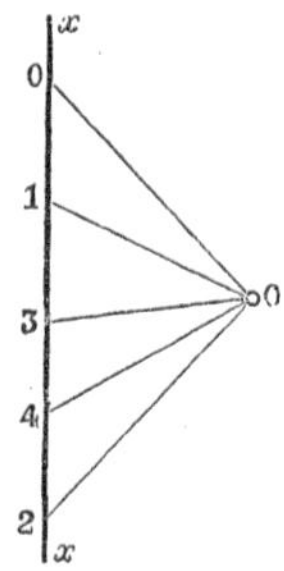

Fig. 54 a.

gonal contour; finally, cut this second pencil by a straight line x, parallel to a, and at a distance b from O measured in the direction of the c's. The segments 01, 12, 23... which we thus obtain on x are those required.

The first and last sides of the polygonal circuit intersect in a point whose distance from a in the direction of the c's we shall call d, and then we shall have $\frac{\Sigma a}{d} = \Sigma \frac{a}{c}$. For it is clear that $\Sigma \frac{a}{c} = \frac{\Sigma x}{b}$, and from the two similar triangles, one of which is bounded by a and the first and last sides of the polygonal circuit, and the other by x and the first and last rays proceeding from O, we have $\frac{\Sigma a}{d} = \frac{\Sigma x}{b}$.

This problem is substantially the same as that of transforming a number of given fractions,

$$\frac{a_1}{c_1}, \quad \frac{a_2}{c_2}, \text{ \&c.,}$$

into equivalent ones $= \frac{x_1}{b}, \frac{x_2}{b}$, &c., with a common denominator b.

68. PROBLEM. *To multiply a straight line a by the ratios*

$$\frac{b_1}{c_1}, \frac{b_2}{c_2}, \ldots, \frac{b_n}{c_n}.$$

Draw (Fig. 55) two straight lines or axes bb, cc which cut one another, at any angle whatever, in the point O.

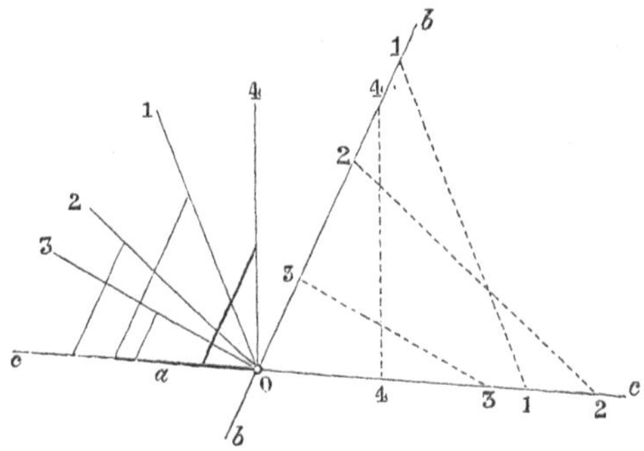

Fig. 55.

From O set off along the first axis the segments b, and along the second the segments c, so that we have on the first axis bb: $O1 = b_1$, $O2 = b_2$ ......, $On = b_n$; and on the second axis cc: $O1 = c_1$, $O2 = c_2$ ......, $On = c_n$. Join the homonymous points of the two axes, i.e. 1 and 1, 2 and 2, and so on, and parallel to the joining lines draw through O the same number of straight lines l_1, l_2, l_3 &c. (which are only denoted in the figure by the numerical index). Two segments b_r, c_r with the same index, and the line rr which joins their final points, form a triangle. Construct a triangle similar to this, in which the two sides corresponding to c_r and rr are set off from O along cc and l_r respectively; the third side corresponding to b_r and parallel to bb, is called a_r. In order to completely determine this triangle, we need only fix one side, that lying on cc; this is equal to a in the first triangle, a_1 in the second, a_2 in the third, and a_{n-1} in the last. Then a_n, that is, that side of the last triangle which is parallel to bb, is the result of the multiplication we wished to perform.

For comparing the rth triangle of the second set, of which the sides parallel to cc and bb are a_{r-1}, a_r, with the similar triangle of the first set, whose corresponding sides are c_r and b_r, we have: $\frac{a_r}{a_{r-1}} = \frac{b_r}{c_r}$; and therefore

$$\frac{a_1}{a} = \frac{b_1}{c_1}, \quad \frac{a_2}{a_1} = \frac{b_2}{c_2}, \ldots, \quad \frac{a_n}{a_{n-1}} = \frac{b_n}{c_n}.$$

Multiply all these equations together, and we have

$$a_n = a.\frac{b_1}{c_1}.\frac{b_2}{c_2}\ldots\frac{b_n}{c_n}.$$

Q. E. D.

69. We shall now prove that the result is not altered by the interchange of two factors, for example, $\frac{b_1}{c_1}$ and $\frac{b_2}{c_2}$. Taking them in the order $\frac{b_1}{c_1}$ and $\frac{b_2}{c_2}$, the construction is as follows (Fig. 56): on cc take $OA = a$; from A draw a parallel to bb, cutting l_1 in A_1; the segment $AA_1 = a_1$, carry this over on to cc, i.e. set off $OA_1 = a_1$, and from this new point A_1 draw a parallel to bb, cutting l_2 in A_2; the segment $A_1\,A_2$ thus obtained is a_2. Now take the factors in the other order $\frac{b_2}{c_2}$ and $\frac{b_1}{c_1}$; and proceed as follows:

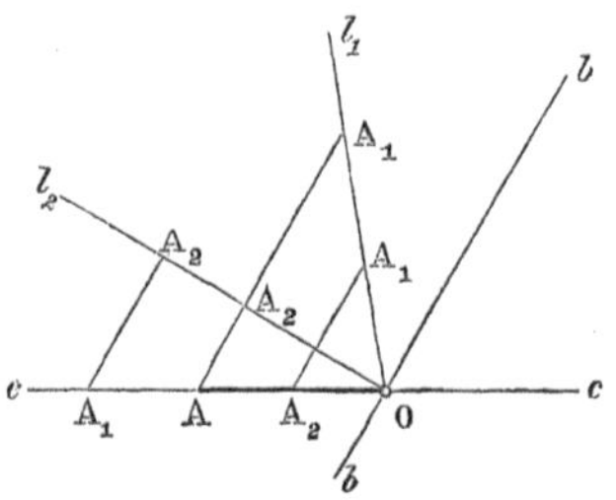

Fig. 56.

Make $OA = a$ as before, and draw through A a parallel to bb, let it cut l_2 in the point A_2, and call the segment so obtained a'; then set off on cc the segment $OA_2 = a'$, and draw $A_2\,A_1$ parallel to bb cutting l_1 in A_1; the straight line $A_2\,A_1$ is a''. The similar triangles contained between l_1 and cc, $OA_1\,A_2$, $OA_1\,A$ give the following relations,

$$\frac{A_2A_1}{AA_1} = \frac{OA_2}{OA}, \text{ i.e. } \frac{a''}{a_1} = \frac{a'}{a};$$

and the similar triangles OA_1A_2, OAA_2 lying between l_2 and cc, give in like manner

$$\frac{A_1A_2}{OA_1} = \frac{AA_2}{OA}, \text{ or } \frac{a_2}{a_1} = \frac{a'}{a}.$$

And therefore $a'' = a_2$. Q. E. D.*

* Eggers, *Grundzüge einer graphischen Arithmetik* (Schaffhausen, 1865), p. 12. Jaeger, l. c., p. 11.

70. In constructing the triangles of the first set, instead of setting off the segments b on the straight line bb, we might, after taking on cc the side $O1 = c_1$, find on Ob a point 1, such that the joining line 11 would be equal in absolute length to b_1. Then having drawn through O, l_1 parallel to 11 we might, as above, construct a triangle of the second set, similar to $O11$, setting off on cc a side equal to a. Then the product (Fig. 57) $a_1 = a \cdot \frac{b_1}{c_1}$

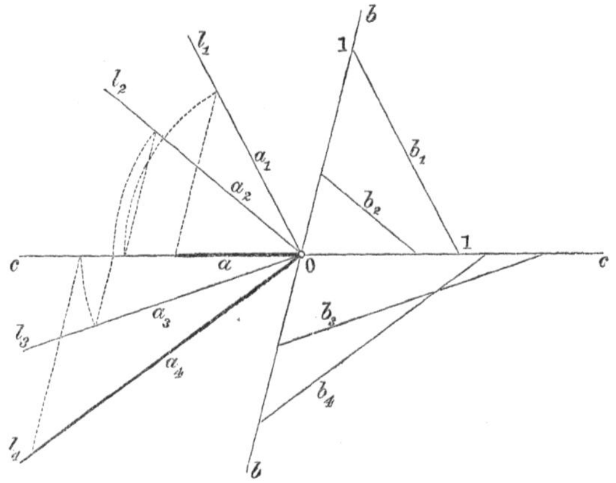

Fig. 57.

is given not by the side parallel to bb, but by the side lying on l_1; and similarly for the other triangles.

In this construction the signs of the segments b are not taken account of, since they are all set off in different directions; it is therefore necessary in carrying over the segments, for example a_1, upon cc, in order to proceed with the construction of the next triangle in the series, to give a_1 the same sign as a, or the opposite according as b_1 and c_1 have the same or opposite signs.

In this method the segments $a_1, a_2, \ldots, a_n$, which we have respectively obtained on $l_1, l_2, \ldots, l_n$ (the parallels to $b_1, b_2, \ldots, b_n$), are carried over to cc by means of circular arcs described around O as centre.

71. A third method of performing the required multiplication is as follows. Set off from the common point O along one of the two axes (bb) the segments b_1, b_3, b_5, &c., and c_2, c_4, c_6, &c.; and along the other axis (cc) (Fig. 58), b_2, b_4, &c. and c_1, c_3, c_5, &c., always joining the extremities 11, 22, 33, &c.

of the segments b and c with the same index. Then it is only necessary to inscribe between the two axes a crooked line whose successive sides are respectively parallel to the joining lines 11, 22, &c., and whose vertices lie alternately on cc and bb.

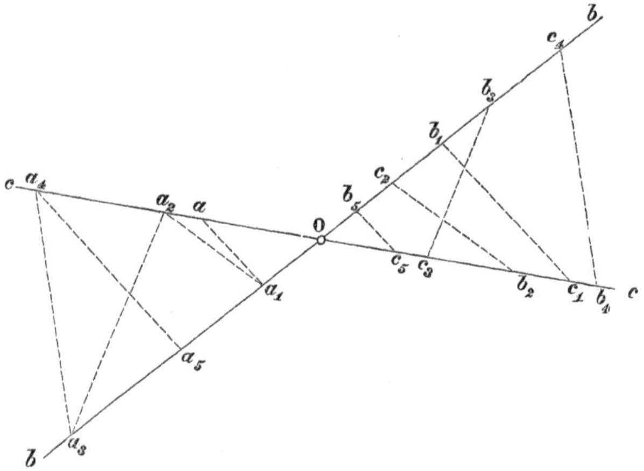

Fig. 58.

If we take the first vertex, so that it is the final point of that segment of cc which is equal to a, and has its initial point at O, then the second vertex, and the third, fourth, &c., are likewise the final points of the segments

$$a_1 = a \cdot \frac{b_1}{c_1}, \quad a_2 = a_1 \cdot \frac{b_2}{c_2}, \quad a_3 = a_2 \cdot \frac{b_3}{c_3} \text{ \&c.,}$$

whose common initial point is O*.

This is evident, when we consider that in this construction all the triangles of the second set have one side on cc and the other on bb, whilst the third side, on the crooked line, is parallel to the third side of the similar triangle of the first set.

72. When there is no need to take account of the signs of the segments a, b, c, i. e. when they may all be considered positive, we can so order the construction, that both the triangles of the first and of the second sets are placed consecutively around a common vertex O (Fig. 59) (just like a

* In Figs. 58, 59, and the following, each of the segments whose common initial point is O, is marked at its final point with the letter a, b, or c, which indicates its measure.

fan). Through O draw $n+1$ straight lines, or radii vectores, making arbitrary angles with one another. Between the first and second radii construct the first triangle of the first set, and the first of the second; between the second and the third radius vector the second triangles of both sets; between the third and the fourth radius vector the third triangles; and so on; in such a manner that two consecutive triangles of the second set always have one side in common. That is to say, starting from O, take on the first radius two segments equal to a and c_1 respectively; on the second radius a segment, with the same initial point, equal to b_1; join the final points of b_1, c_1, and draw a parallel to the joining line through the final point of the segment a, this determines a segment a_1 on the second radius, such that

$$a_1 = a\frac{b_1}{c_1}.$$

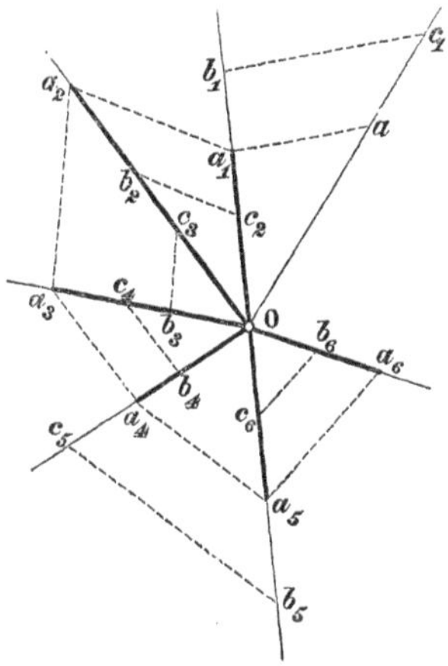

Fig. 59.

Now take, in the same way, the segment c_2 on the second radius, and the segment b_2 on the third radius, and we determine on the latter a segment

$$a_2 = a_1 \cdot \frac{b_2}{c_2} = a\frac{b_1}{c_1} \cdot \frac{b_2}{c_2}. \text{ \&c.}$$

Continuing this construction, we finally get, on the $(n+1)^{\text{th}}$ radius, a segment with its initial point at O, whose value will be

$$a_n = a \cdot \frac{b_1}{c_1} \cdot \frac{b_2}{c_2} \cdot \ldots \cdot \frac{b_n}{c_n}.$$

CHAPTER IV.

POWERS.

73. If, in the last problem, we make all the b's equal to one another, as also all the c's, then the constructed segment a_n is the result of multiplying a by the nth power of the ratio $\frac{b}{c}$.

In this case, either in the first construction Art. 68 (Fig. 60),

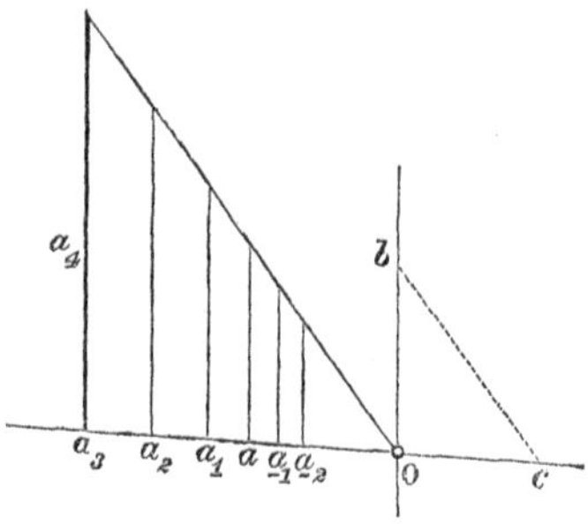

Fig. 60.

or in the second, Art. 70 (Fig. 61), all the triangles of the first set coincide, and form a single triangle, two of whose

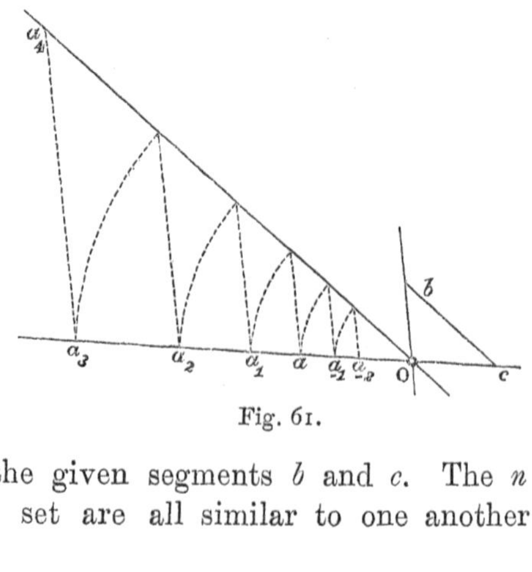

Fig. 61.

sides are the given segments b and c. The n triangles of the second set are all similar to one another and to the single triangle Obc. The sides lying on Oc are respectively $a, a_1, a_2, \ldots, a_{n-1}$, whilst the sides parallel to b are $a_1, a_2, \ldots, a_n$, and therefore

$$a_1 = a\frac{b}{c}, \quad a_2 = a\left(\frac{b}{c}\right)^2, \quad a_3 = a\left(\frac{b}{c}\right)^3, \ldots, \quad a_n = a\left(\frac{b}{c}\right)^n.$$

This series of similar triangles can also be prolonged on the opposite side, so as to give the product of a by the negative powers of $\frac{b}{c}$. In fact, constructing the triangle whose side parallel to b is equal to a, the side which lies upon Oc is

$$a_{-1} = a \cdot \frac{b}{c} = a \cdot \left(\frac{b}{c}\right)^{-1};$$

next, constructing a triangle with its side parallel to $b = a_{-1}$, the side on $OC = a_{-2} = a\left(\frac{b}{c}\right)^{-2}$, and so on*.

74. In the third method (Art. 71) the triangles of the first set reduce to two equal, but differently situated, triangles Obc (Fig. 62); the one has its side c on the first axis and its

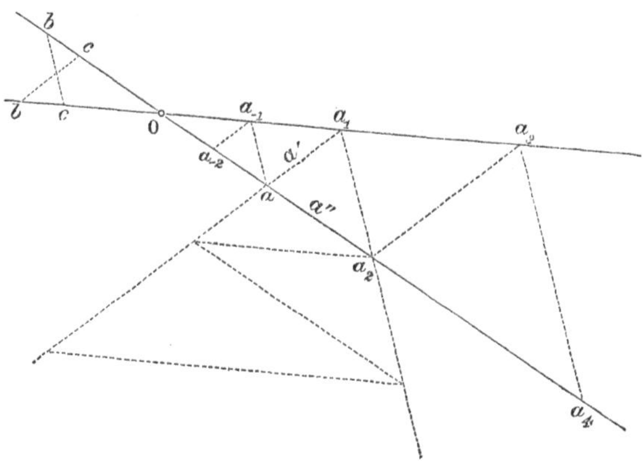

Fig. 62.

side b on the second; whilst the other has its side b on the first axis and c on the second. The directions of the third sides are therefore antiparallel, and the sides of the crooked line, inserted between the two axes, are parallel to them.

* EGGERS, l. c., p. 15. JAEGER, l. c., pp. 18–20.

The vertices of this crooked line determine on the first axis segments, measured from O, which have the values

$$a, \quad a_2 = a\left(\frac{b}{c}\right)^2, \quad a_4 = a\left(\frac{b}{c}\right)^4, \ldots,$$

and on the other axis

$$a_1 = a\left(\frac{b}{c}\right), \quad a_3 = a\left(\frac{b}{c}\right)^3, \text{ \&c.*}$$

Moreover the sides of the crooked line form a geometrical progression; for, if we call the first side a', the second is $a'\frac{b}{c}$, the third $a'\left(\frac{b}{c}\right)^2$, the fourth $a'\left(\frac{b}{c}\right)^3$, &c.

Hence we conclude that the given segment, which has to be multiplied by $\left(\frac{b}{c}\right)$, instead of being set off on the first axis, may be placed in the angle between the axes so as to form the first side of the crooked line; its $(n+1)^{\text{th}}$ side will then be the result of the multiplication.

Continuing the crooked line in the opposite direction we obtain the products of the given segment (a or a') by the negative powers of the given ratio

$$\left(\frac{b}{c}\right)^{-1}, \quad \left(\frac{b}{c}\right)^{-2}, \text{ \&c.}$$

If we wish to continue the progression between two successive sides of the crooked line, for example, between the two first $\left(a' \text{ and } a'.\frac{b}{c}\right)$, then we need only draw between them a new crooked line, whose sides are alternately parallel to the axes; and we obtain a figure analogous to the foregoing one.

Let the segment of the first axis, which is intercepted by the first two sides of the first crooked line, be called a'', then the sides of the new crooked line are respectively

$$a'', \quad a''\frac{b}{c}, \quad a''\left(\frac{b}{c}\right)^2, \text{ \&c.†}$$

75. Finally, if we employ the fourth method of construction

* COUSINERY, l. c., p. 24, 25.
† COUSINERY, l. c., p. 24. CULMANN, l. c., No. 3.

(Art. 72) and take the angle between consecutive radii vectores constant (Fig. 63), all the triangles of the first series become equal and their vertices (opposite O) lie on two concentric circles whose radii are respectively equal to b and c. The triangles of the second series are all similar to one another, because each is similar to the corresponding triangle of

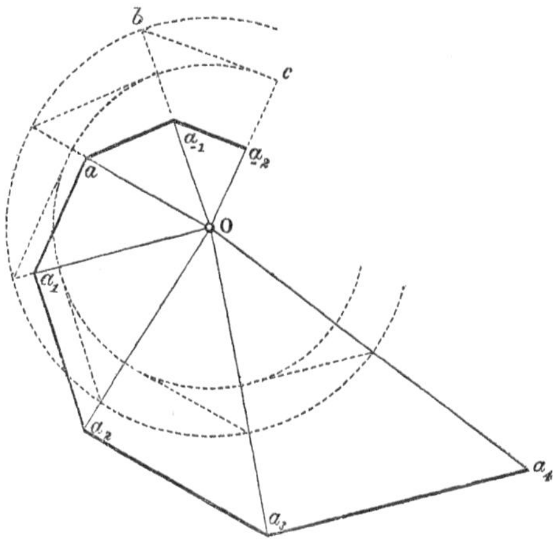

Fig. 63.

the first series; their vertices (opposite O), and their sides (lying opposite O) are the vertices and sides of a polygonal spiral circuit.

The radii vectores of this spiral, i.e. the straight lines drawn from O to the vertices, are the terms of a geometrical progression

$$a, \quad a_1 = a\frac{b}{c}, \quad a_2 = a\left(\frac{b}{c}\right)^2, \text{ \&c.}$$

This progression may be continued in the opposite direction, so as to give the products of a by the negative powers of $\left(\frac{b}{c}\right)$:

$$a\left(\frac{b}{c}\right)^{-1}, \quad a\left(\frac{b}{c}\right)^{-2} \text{ \&c., \&c.}$$

Also the sides of the polygonal circuit form a geometrical progression with the same common ratio $\frac{b}{c}$ *.

* Jaeger, l. c., p. 20.

If the constant angle between two consecutive radii vectores is a commensurable fraction of four right angles, which has the denominator p when reduced to its lowest terms, then the

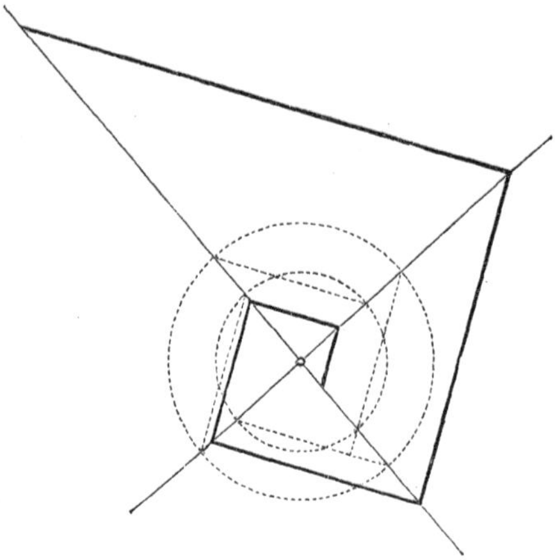

Fig. 64.

$(p+1)^{th}$ radius coincides with the first, the $(p+2)^{th}$ with the second, and so on. If, for example, the given constant angle were a right angle*; the angles between every pair of consecutive sides of the spiral polygon would also be right angles (Fig. 64).

* Reuleaux, *Der Constructeur*, 3rd edition (Braunschweig, 1869), p. 84. K. von Ott, *Grundzüge des graphischen Rechnens und der graphischen Statik*, (Prag. 1871), p. 10.

CHAPTER V.

EXTRACTION OF ROOTS.

76. CONSIDER the spiral polygon $ABCDEFG \ldots$ (Fig. 65), whose radii vectores OA, OB, OC, OD, &c. represent the products of a constant segment OA by the powers (corresponding to the indices 0, 1, 2, 3, &c.) of a given ratio $\frac{b}{c} = \frac{OB}{OA}$, and whose sides AB, BC, CD, &c. subtend a constant angle at the

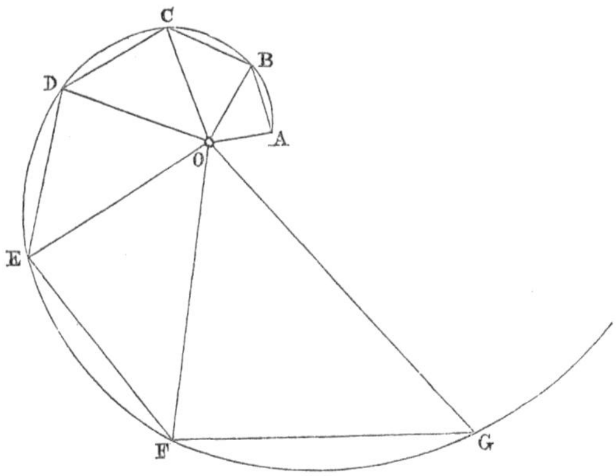

Fig. 65.

pole O (Art. 75). As already remarked, all the elementary triangles, which have O for a vertex and a side of the polygon as base, are similar; also all the figures, obtained by combining 2, 3 or 4, &c. consecutive triangles, are similar, because they are made up of the same number of similar and similarly situated triangles. Therefore all the angles ABO, BCO, CDO, &c. are equal; also the angles ACO, BDO, CEO, &c.; and so on. In general all the triangles around the vertex O, the bases of which are chords, joining the extreme points of the same number of consecutive sides of the polygon, are similar; these chords also subtend equal angles at the pole O.

These properties are quite independent of the magnitude

of the angle AOB, which in the construction of the first elementary triangle is chosen at pleasure. They would not therefore cease to be true if this angle were made infinitely small: in which case the polygonal circuit becomes a curve. From the similarity of the elementary triangles we have already deduced the equality of the angles at the bases OAB, OBC, &c.; but if the angles at the point O become infinitely small, the sides of the elementary triangles lying opposite to O will become tangents to the curve; the curve obtained has therefore the property, that its tangents (produced in the same sense, for example, in that of the increasing radii vectores) meet* the radii vectores, drawn from the pole O to the point of contact, at equal angles. From this property this curve is called *The Equiangular Spiral* †.

77. Since the figures, which are made up of an equal number of successive elementary triangles, are similar, so also, if we draw in the equiangular spiral the radii vectores OA, OB, OC, &c., at equal angular intervals, the triangles OAB, OBC, OCD, &c., will be similar to one another. Therefore the radii vectores in question form a geometrical progression, i.e. the polygonal circuit $ABCD$... inscribed in the spiral is exactly the same as the one constructed by the rule of Art. 75, starting from the elementary triangle AOB. If therefore we take the triangle AOB at pleasure, and construct the polygonal circuit $ABCD$..., all its vertices lie on the same equiangular spiral with its pole at O. Hence it follows, that the pole and two points of the curve completely determine an equiangular spiral.

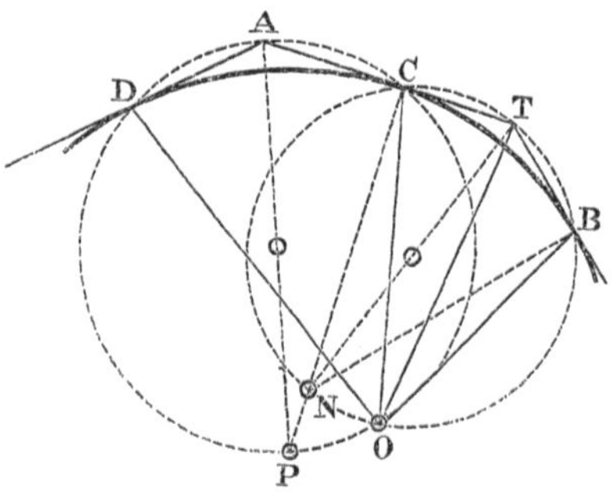

Fig. 66.

78. Any two points B, C (Fig. 66) of an equiangular spiral,

* COUSINERY, l. c., p. 41, 42. CULMANN, l. c., No. 5.

† WHITWORTH, *The equiangular spiral, its chief properties proved geometrically* (Oxford, Cambridge, and Dublin Messenger of Mathematics, vol. i. p. 5, Cambridge, 1862).

the pole O, the point of intersection T of the tangents at those points, and the point of intersection N of the corresponding normals, are five points on the circle whose diameter is NT. Of the truth of this we are easily convinced if we consider, (1) that the circle drawn on NT as a diameter will pass through the points B and C, since the angles NBT and NCT are right angles; (2) that the angles OBT and OCT being supplementary (since the angle made by a tangent with the radius vector drawn to its point of contact is constant), the four points $OTBC$ belong to the same circle. Hence it follows that NOT is a right angle.

79. Now take the points B and C so close together, that the spiral arc between them can be replaced by a circular arc. Since this arc must touch BT and CT in the points B and C, its centre lies at N; the tangents BT, CT are equal, and therefore the chord BC is bisected at right angles by the straight line NT; hence also, N and T are the points of bisection of the arcs BC of the circle OBC, i.e. OT is the internal, and ON the external bisector of the angle BOC. The point N, which will serve as a centre from which to describe the arc BC substituted for the spiral arc, can therefore be constructed as the extremity of that diameter of the circle OBC, which is perpendicular to the chord BC. The centre P of the next arc CD, which must be the point of intersection of the normals at C and D, will be the point of intersection of the straight line CN, with the straight line which bisects the chord CD at right angles, or with the external bisector of the angle COD. And so on.

80. From this we obtain a construction for the equiangular spiral by means of circular arcs. We divide (Fig. 67) the angular space (four right angles) round the pole O into a certain number of equal parts, so small that the spiral arc corresponding to each part can be replaced by a circular arc. On two consecutive radii vectores points A and B are taken, through which the spiral must pass. The centre M for the arc AB is then the end of that diameter of the circle OAB, which is at right angles to the chord AB. Let N be the point where BM cuts the external bisector of the angle between OB and the next radius vector. With the centre N describe the arc BC. Similarly let P be the point, in which CN cuts the external bisector of the angle between OC and the radius vector immediately following it; then with P as centre we describe the arc CD; and so on*.

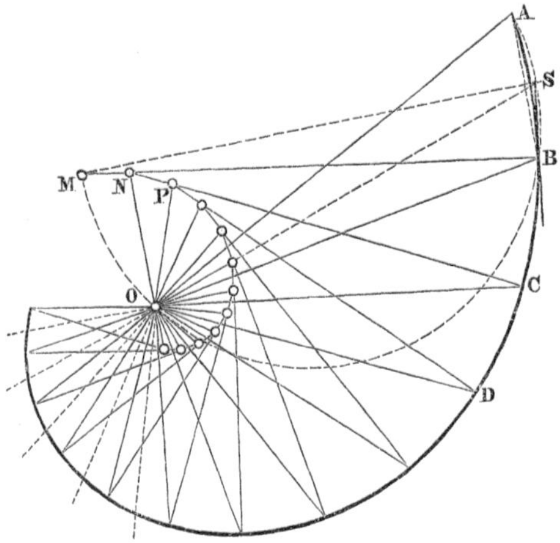

Fig. 67.

81. Instead of assuming the point A (as well as O and B), we may suppose the constant angle between the tangent and radius vector to be given. In this case, having drawn BS inclined to OB at the given angle, let S be the point of intersection of this tangent BS with the internal bisector of the angle between OB and its preceding radius vector; then the point A is given by the intersection of that radius vector with the circle OBS. After we have found the point M of this circle, which is diametrically opposite to S, we proceed with the construction in the manner explained above†.

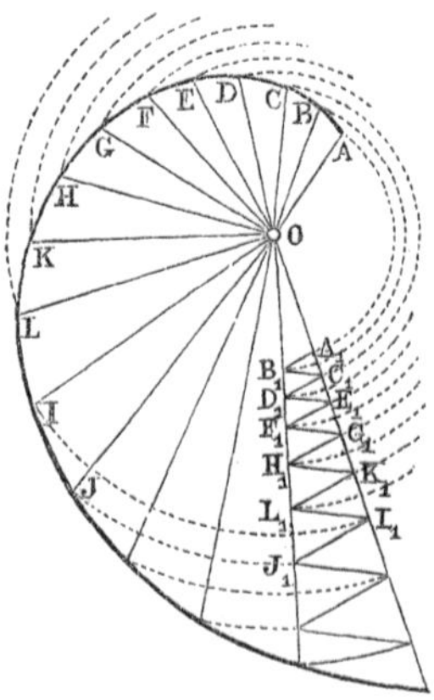

Fig. 68.

82. We are often able to avoid drawing these circular arcs, and to restrict ourselves to finding a series of points on the curve sufficiently near together to be united to one another by a continuous line. For this purpose we take the

* For this construction I am indebted to Prof. A. Sayno, of Milan.

† For this construction I am also indebted to Prof. A. Sayno.

elementary triangle OA_1B_1 (Fig. 68), of which the angle at O is very small, and between the sides OA_1 and OB_1 construct the crooked line $A_1\,B_1\,C_1\,D_1\ldots$, with its sides alternately parallel and antiparallel to $A_1\,B_1$. Then, upon the radii vectores OA, OB, OC, &c., drawn at angular intervals each equal to the constant angle $A_1\,O\,B_1$, take points A, B, C, &c., in such a manner, that $OA_1 = OA$, $OB_1 = OB$, &c.

83. This spiral when drawn serves for the solution of problems involving the extraction of roots.

We want, say, the ith root of the ratio between two given segments a_i, a. Write $a_i = a\left(\frac{b}{c}\right)^i$, then the question becomes that of finding the ratio $\frac{b}{c}$. Take on the spiral (Fig. 69, where $i = 5$) the radii vectores a and a_i, and divide the angle included between them into i equal parts. The $i-1$ dividing radii vectores a_1, a_2, &c., will be the intermediate terms of a geometrical progression of $i+1$ terms, the first of which is a and the last a_i. The ratio $a_1 : a$ of the two first terms is therefore the required ratio $\left(\frac{b}{c}\right)$.

84. Two radii vectores containing a constant angle have a constant ratio. From this it follows, that, if we take the sum or difference of the angles contained by two pairs of radii vectores $a_1\,b_1$ and $a_2\,b_2$, the resulting angle is contained by two radii vectores, whose ratio is in the first case equal to the product, in the second to the quotient, of the ratios $a_1 : b_1$, and $a_2 : b_2$. That is to say, the equiangular spiral renders the same service in graphical calculations which a table of logarithms does in numerical methods. The ratios of the radii vectores correspond to the numbers, the angles to their logarithms.

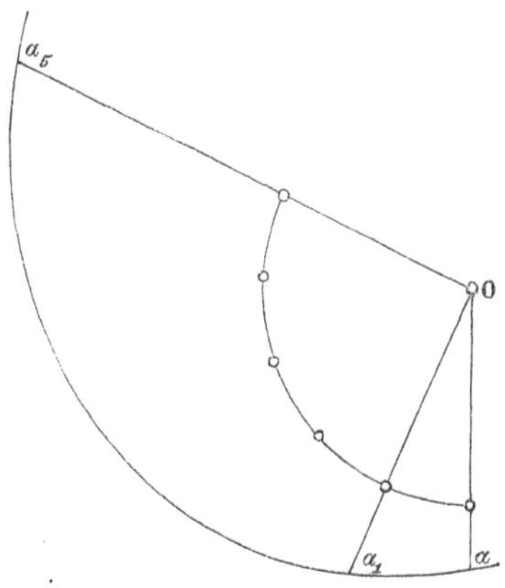

Fig. 69.

On account of this property the curve we are speaking of is also called the *Logarithmic Spiral*. If we take a radius vector

equal to the linear unit as the common denominator of these ratios, it is obvious that the radii vectores themselves may be considered instead of their ratios to unity.

If, for example, we wish to construct the segment x, given by the equation

$$x = \sqrt[n]{a_1 . a_2 . \dots . a_n},$$

then x is the radius vector of the spiral, which makes with the radius 1 an angle equal to the arithmetic mean of the angles, which the radii $a_1, a_2, \dots a_n$ make with the same radius 1.

85. But when the extraction of a square root only is wanted, instead of employing the spiral, it is much easier to use the known constructions of elementary geometry. If, for example, $x = \sqrt{ab}$, we construct x as the geometric mean of the segments a and b.

If the segments $OA = a$, $OB = b$ are set off on a straight line in the same sense, then (Fig. 70) x is the length of the tangent OX, drawn from O to a circle described through A and B; or (Fig. 70a) a circle may be drawn with diameter $= OA$ (the greater segment), and then x is the chord OX, whose projection on the diameter is the other segment b.

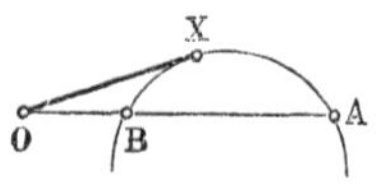

Fig. 70.

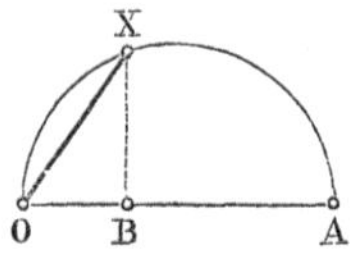

Fig. 70 a.

Again, if the segments $OA = a$, $OB = b$ lie in a straight line, but have opposite sense (Fig. 71), we describe a semicircle on AB, and then x is the ordinate erected at the point O.

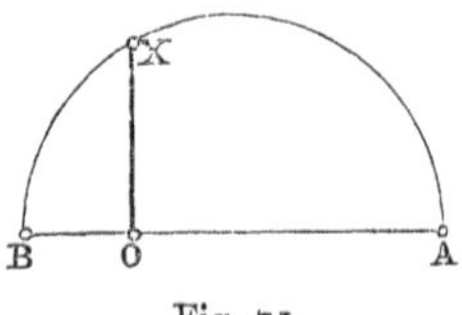

Fig. 71.

86. The same ends for which the equiangular spiral serves, are easily attained by using another curve called '*The Logarithmic Curve.*'

Draw (Fig. 72) two axes Ox and Oy; and on the first of them, starting from the origin O, take the segments $O0$, $O1$, $O2$, $O3$, &c., respectively equal to the terms

$$x_0,\ x_1 = x_0 \frac{m}{n},\ x_2 = x_0 \left(\frac{m}{n}\right)^2,\ x_3 = x_0 \left(\frac{m}{n}\right)^3, \text{\&c.},$$

of a geometrical progression, of which the first term is x_0, and the common ratio $\frac{m}{n}$ (where m is supposed greater than n); and on the second axis take the segments $O0$, $O1$, $O2$, $O3$, &c., also measured from O, and respectively equal to the terms $y_0 = 0$, $y_1 = l$, $y_2 = 2l$, $y_3 = 3l$, &c., of an arithmetical progression, with its first term equal to zero, and the common difference* $= l$. The terms of the two progressions, which correspond to the index r, are

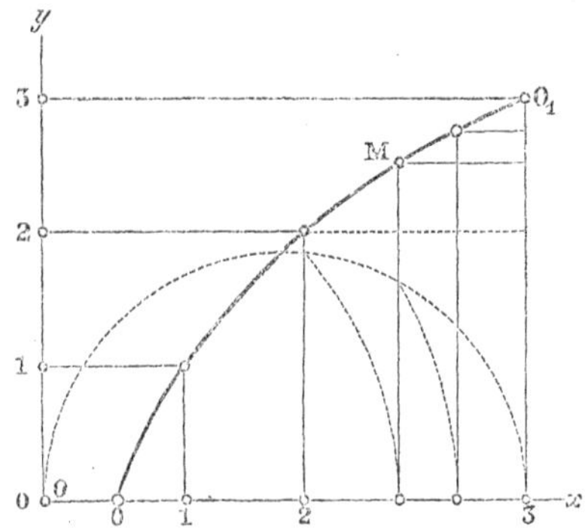

Fig. 72.

$$x_r = x_0 \left(\frac{m}{n}\right)^r, \quad y_r = rl;$$

and therefore
$$x_r = x_0 \left(\frac{m}{n}\right)^{\frac{y_r}{l}}.$$

Between each pair of consecutive terms in each of the two progressions we can interpolate a new term, so as to obtain two new progressions, of which the first has a common ratio $\left(\frac{m}{n}\right)^{\frac{1}{2}}$ or $\frac{\sqrt{mn}}{n}$, and the other a common difference $\frac{l}{2}$. This follows, from the fact that in every geometrical (arithmetical) progression any term whatever is the geometrical (arithmetical) mean between the terms preceding and following it.

If we construct, for example, the geometrical mean between x_r, and x_{r+1}, and the arithmetical between y_r and y_{r+1}, we obtain the two corresponding terms

$$(x_r\, x_{r+1})^{\frac{1}{2}} = x_0 \left(\frac{m}{n}\right)^{r+\frac{1}{2}}$$
$$\tfrac{1}{2}(y_r + y_{r+1}) = (r + \tfrac{1}{2})\, l$$

of the two new progressions.

In these progressions we can in like manner interpolate a

* In the succession of numbers on Oy, the zero coincides with the origin O, because y_0 was taken $= 0$.

term between each pair of consecutive terms, and so on, until we arrive at two progressions, for which the ratio $\left(\frac{m}{n}\right)^{\frac{1}{2^i}}$ and the difference $\frac{l}{2^i}$ are as small, as we please*. If we use x and y to denote two corresponding terms, we have always

$$(1)\quad x = x_0\left(\frac{m}{n}\right)^{\frac{y}{l}},$$

or

$$(2)\quad y = l\,\frac{\log\frac{x}{x_0}}{\log\frac{m}{n}},$$

the logarithms being taken in any system whatever. We shall call those points of the axes Ox, Oy *corresponding points*, in which corresponding segments x and y terminate. We draw parallels to the axes through these corresponding points, i.e. through the final point of x a parallel to Oy, and through the final point of y a parallel to Ox. The straight lines so drawn will intersect in a point M; x and y are then called the co-ordinates of the point M, and in particular x is called the '*abscissa*' and y the '*ordinate*.' The equation (1) or (2) expressing the relation between the co-ordinates of the point M, is called the *equation* to the curve which is the locus of all points analogous to M. We call this curve the '*Logarithmic Curve*' because the ordinate is proportional to the logarithm of a number which is proportional to the abscissa.

87. We construct this curve 'by points' in the following manner. After drawing the two axes Ox, Oy (Fig. 73) (usually at right angles) we take on Oy a segment $OB = O\,(2^i) = l$, where l may be considered as the unit of the scale of lengths on Oy; and, upon Ox we take a segment $OA = O\,(2^i) = x_0\,\frac{m}{n}$, where $O0 = x_0$ is the unit of length of the scale for Ox†, and $\frac{m}{n}$ the base of the logarithmic system (the number 10).

* i is the number of interpolations.

† Since x increases much faster than y, it is convenient, in order to keep the construction within reasonable limits, to take the unit x_0 much smaller than l, for ex.,

$$x_0\frac{m}{n} = l.$$

Let OB be divided into 2^i equal parts, and let $1, 2, 3, \ldots, 2^{i-1}, \ldots, 2^i (= B)$ be the points of division.

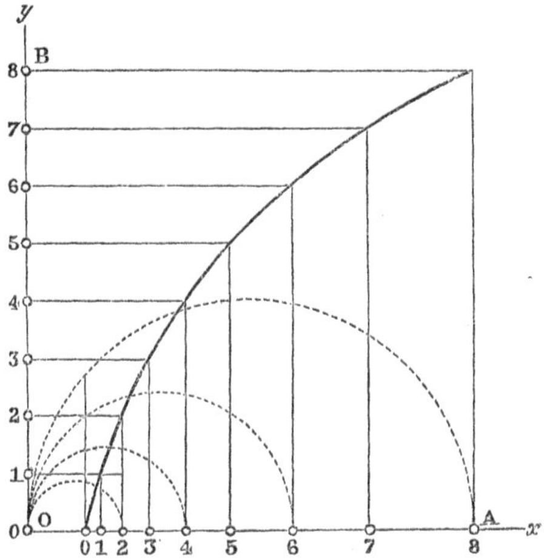

Fig. 73.

In order to find the corresponding points of Ox, take the geometrical mean between x_0 and $x_0\frac{m}{n}$, i.e. describe a semicircle on OA as diameter, and set off along OA, starting from O, the length of the chord of this semicircle, which has $O0$ for a projection; we shall thus obtain the point 2^{i-1} of Ox, which corresponds to the similarly named point of Oy (i.e. to the middle point of OB). Similarly by taking the geometrical mean of $O0$ and $O2^{i-1}$, and the geometrical mean between $O2^{i-1}$ and OA, we shall obtain the points on Ox, corresponding to the middle points of the segments $O2^{i-1}$, and $2^{i-1}B$ of Oy; and so on.

Now draw parallels to Oy through the points of division of Ox, and parallels to Ox through the points of division of Oy; the points, in which the lines drawn through similarly named points intersect, lie on the logarithmic curve we are constructing. Since to the value $y = y_0 = 0$ corresponds the value $x = x_0 = O0$, the curve passes through the point marked 0 on Ox.

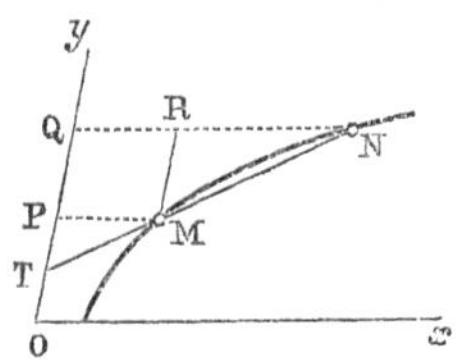

Fig. 74.

88. It is also very easy to construct a tangent to the curve at any one of its points (Fig. 74). Let M and N be any two points on the curve, at a small distance from one another; MP, NQ parallels to Ox, MR a parallel to Oy, and T the point in which Oy is cut by the chord MN. The similar triangles TPM and MRN give

$$TP : MP = MR : NR,$$

or $$TP : MP = OQ - OP : NQ - MP.$$

Let $OP = y$, $PQ = h$, then MP and NQ are the abscissae corresponding to the ordinates y, $y+h$, and therefore

$$MP = x_0 \left(\frac{m}{n}\right)^{\frac{y}{l}}, \quad NQ = x_0 \left(\frac{m}{n}\right)^{\frac{y+h}{l}},$$

whence $$TP = \frac{h\left(\frac{m}{n}\right)^{\frac{y}{l}}}{\left(\frac{m}{n}\right)^{\frac{y+h}{l}} - \left(\frac{m}{n}\right)^{\frac{y}{l}}} = l \frac{\frac{h}{l}}{\left(\frac{m}{n}\right)^{\frac{h}{l}} - 1}.$$

Now let the point N approach continually nearer and nearer to the point M, i.e. until h approximates to the value zero, then NMT will also continually approach towards the position of the tangent at M, and the segment TP, the projection of TM upon Oy, has for its limiting value what is usually called the '*subtangent*.' But the limit * which the fraction

$$\frac{\left(\frac{m}{n}\right)^{\frac{h}{l}} - 1}{\frac{h}{l}}$$

approaches, when h tends towards zero, is the natural logarithm of $\frac{m}{n}$, which we shall denote by $\lambda \frac{m}{n}$; therefore in the limit we have

$$TP = \frac{l}{\lambda \frac{m}{n}},$$

i.e. the subtangent is constant for all the points on the curve †.

Hence it follows, that a single construction suffices for drawing tangents at all the different points on the curve.

89. Having thus constructed the logarithmic curve, we can solve by its aid all those problems for which the ordinary logarithmic tables are used. We want for instance

* BALTZER, *Elemente der Mathematik*, i. 4 ed. (Leipzig, 1872), p. 200.

† SALMON, *Higher plane curves*, 2nd ed. (Dublin, 1873), p. 314.

to construct the rth root of the ratio between two straight lines p, q. Take upon Ox the abscissae $x' = p$, $x'' = q$, and find by means of the curve the corresponding ordinates y' and y''.

The abscissa corresponding to the ordinate $\frac{1}{r}(y' - y'')$ has the value

$$x_0 \sqrt[r]{\frac{p}{q}}.$$

Secondly, say we want the rth root of the product of the r straight lines $p_1, p_2, \ldots, p_r$. Take on Ox the abscissae $x_1 = p_1$, $x_2 = p_2$, &c., and find the corresponding ordinates $y_1, y_2, y_3, \ldots y_r$; then the abscissa x corresponding to the ordinate

$$\frac{1}{r}(y_1 + y_2 + \ldots + y_r)$$

is equal to the required quantity

$$\sqrt[r]{p_1, p_2, p_3, \ldots p_r}.$$

CHAPTER VI.

SOLUTION OF NUMERICAL EQUATIONS*.

90. LET $a_0, a_1, a_2, \ldots, a_n$ be $n+1$ numbers given in magnitude and sign, and let (Fig. 75) a polygonal right-angled circuit be constructed, the lengths of whose successive sides 01, 12, 23, ... are proportional to the given numbers. The sense of each side is determined by the following law: the rth and the $(r+2)^{\text{th}}$ sides, which are parallel to one another, have the same or opposite sense, according as the signs of the numbers a_{r-1}, a_{r+1}, which are proportional to these sides, are unlike or alike†.

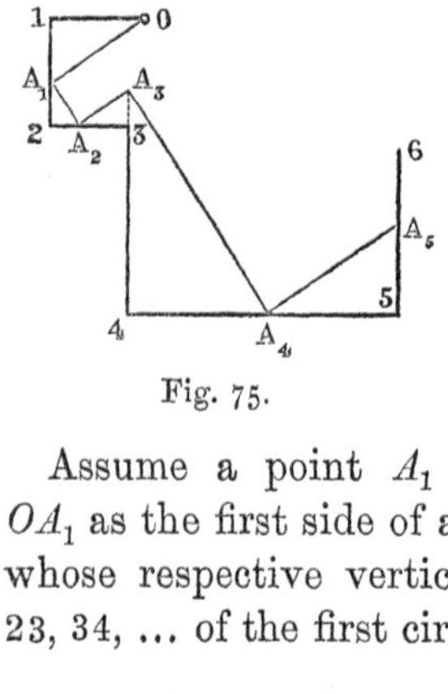

Fig. 75.

Assume a point A_1 in the straight line 12, and take OA_1 as the first side of a second right-angled circuit of n sides, whose respective vertices $A_1, A_2, A_3, \ldots$ lie on the sides 12, 23, 34, ... of the first circuit.

* LILL, *Résolution graphique des équations numérique d'un degré quelconque à une inconnue* (Nouvelles Annales de Mathématiques, 2e série, t. 6, Paris, 1867), p. 359.

† In order to fix with the greatest possible precision the sense of each side of the crooked line, the following convention is useful. We take two rectangular axes XOX, YOY, and determine for each of them the positive sense; and we agree to give the number, which expresses the length of a segment, the coefficient $+1$, or -1, according as it is in the positive or negative direction of XOX, and the coefficient $+i$, or $-i$ (where $i = \sqrt{-1}$, i. e., $i^2 = -1$), according as it is in the positive or negative direction of YOY. Now let a circuit be formed whose successive sides,

$$01,\ 12,\ 23,\ 34,\ 45,\ 56,\ \ldots.$$

are equal to $\qquad a_0,\ ia_1,\ i^2a_2\ i^3a_3,\ i^4a_4,\ i^5a_5, \ldots,$

i.e. equal to $\qquad a_0,\ ia_1,\ -a_2,\ -ia_3,\ a_4,\ ia_5, \ldots,$

then the 1st, 3rd, 5th, ... sides will be parallel to XOX, and the others to YOY; moreover two parallel sides, separated by a single side at right angles to both, will have the same, or opposite sense according as the corresponding numbers a have opposite signs or the same sign.

The triangles $01A_1$, $A_1 2 A_2$, $A_2 3 A_3$, $A_3 4 A_4$, ... are all similar to one another, and therefore give

$$\frac{01}{A_1 1} = \frac{A_1 2}{A_2 2} = \frac{A_2 3}{A_3 3} = \frac{A_3 4}{A_4 4} = \dots = \frac{A_{n-1}.n}{A_n.n},$$

whence, remembering the identities,

$$\begin{aligned} 01 &= a_0, & A_1 2 &= A_1 1 + a_1, \\ 12 &= a_1, & A_2 3 &= A_2 2 + a_2, \\ 23 &= a_2, & A_3 4 &= A_3 3 + a_3, \\ \dots & \dots & \dots & \dots \dots \dots \\ n-1, n &= a_{n-1}, & A_n.n+1 &= A_n.n + a_n, \\ n.n+1 &= a_n, \end{aligned}$$

and putting

$$\frac{A_1 1}{01} = x, \text{ or } A_1 1 = a_0 x,$$

we obtain:

$$\begin{aligned} A_1 2 &= a_0 x + a_1, \\ A_2 2 &= a_0 x^2 + a_1 x, \\ A_2 3 &= a_0 x^2 + a_1 x + a_2, \\ A_3 3 &= a_0 x^3 + a_1 x^2 + a_2 x, \\ &\dots \\ A_n.n &= a_0 x^n + a_1 x^{n-1} + \dots + a_{n-1} x, \\ A_n.n+1 &= a_0 x^n + a_1 x^{n-1} + \dots + a_{n-1} x + a_n. \end{aligned}$$

Thus the segment $A_n.n+1$, included between the final points of the first and second polygonal circuits, represents the value which the polynomial

$$F(x) = a_0 x^n + a_1 x^{n-1} + \dots + a_n$$

takes, when we substitute for x the ratio of the segment $A_1 1$ to the segment 01 or a_0. Keeping a_0 positive, the signs of x and a_1 will be equal or opposite, according as $A_1 1$, 12 have the same or opposite sense.

If the final points of the two circuits coincide, we have the identity $F(x) = 0$; and then x is called a *root* of the equation $F(z) = 0$. The real roots of the equation $F(z) = 0$ are therefore the ratios $A_1 1 : 01$, which correspond to those right-angled inscribed circuits whose final points coincide with the point $\overline{n+1}$.

On account of this property we say, that the *circuit* $0123 \dots \overline{n+1}$, *represents the whole polynomial* $F(z)$.

91. If in $0123 \dots \overline{n+1}$ we inscribe a new right-angled circuit,

$0B_1B_2\ldots B_n$, and if we denote the ratio of $B_1 1 : 01$ by y, we shall have in like manner

$$Bn.\overline{n+1} = F(y) = a_0y^n + a_1y^{n-1} + \ldots + a_n.$$

For the coefficients a we substitute their values

$$\begin{aligned}
a_0 &= 01,\\
a_1 &= 12 = A_1 2 - A_1 1 = A_1 2 - 01 . x,\\
a_2 &= 23 = A_2 3 - A_2 2 = A_2 3 - A_1 2 . x,\\
a_3 &= 34 = A_3 4 - A_3 3 = A_3 4 - A_2 3 . x,\\
&\ldots\ldots\ldots\ldots\\
a_{n-1} &= \overline{n-1}.n = A_{n-1}.n - A_{n-1}.\overline{n-1} = A_{n-1}.n - A_{n-2}.\overline{n-1}x,\\
a_n &= n.\overline{n+1} = A_n\overline{n+1} - A_n n = A_n\overline{n+1} - A_{n-1}.n.x;
\end{aligned}$$

and thus obtain

$$\begin{aligned}
B_n.\overline{n+1} = 01.y^n &+ (A_1 2 - 01.x)y^{n-1}\\
&+ (A_2 3 - A_1 2.x)y^{n-2}\\
&+ \ldots\ldots\\
&+ (A_{n-1}.n - A_{n-2}\overline{n-1}x)y\\
&+ (A_n\overline{n+1} - A_{n-1}n.x)
\end{aligned}$$

$$= (y-x)[01.y^{n-1} + A_1 2y^{n-2} + A_2 3y^{n-3} + \ldots + A_{n-1}n] + A_n.\overline{n+1}.$$

But $\quad B_n\overline{n+1} - A_n\overline{n+1} = B_nA_n,$

and $\quad y - x = \dfrac{B_1 1 - A_1 1}{01} = \dfrac{B_1A_1}{01};$

therefore

$$01.\frac{B_nA_n}{B_1A_1} = 01.y^{n-1} + A_1 2.y^{n-2} + A_2 3.y^{n-3} + \ldots + A_{n-1}.n.$$

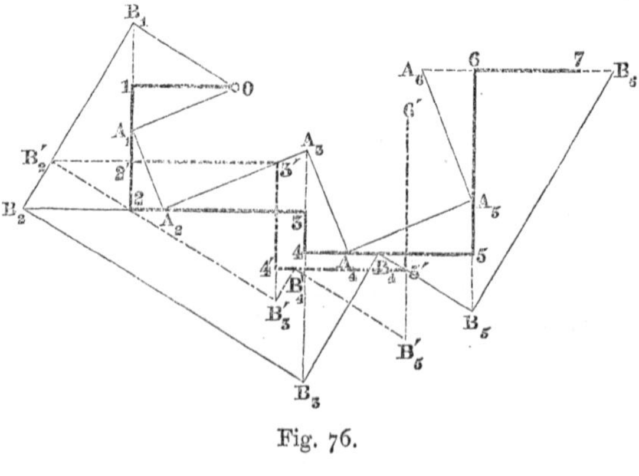

Fig. 76.

This result may be expressed as follows (Fig. 76, where $n = 6$): *In a rectangular circuit of* $n+1$ *sides* $0123\ldots n+1$, *two other rectangular circuits of n sides* $OA_1A_2\ldots A_n$, $OB_1B_2\ldots B_n$ *are inscribed; we then form a new rectangular circuit* $012'3'\ldots n'$ *of n sides, which are respectively parallel to the sides of the first circuit and equal to* 01, $A_1 2$, $A_2 3, \ldots, A_{n-1}n$. *In this inscribe the rectangular circuit of* $n-1$ *sides* $OB_1B_2'\ldots B'_{n-1}$, *having the side* OB_1 *in common with the circuit already described* $OB_1B_2\ldots B_n$.

Then we have
$$\frac{B'_{n-1}n'}{01} = \frac{B_nA_n}{B_1A_1}.$$
That is to say, the segment $B'_{n-1}.n'$ *is the result of the division of* $F(y)-F(x)$ *by* $y-x$: *where F signifies the Polynomial represented by the first circuit* $0123\ldots\overline{n+1}$, *and x and y are the ratios* $A_1 1:01$; $B_1 1:01$.

Or, in other words,

The circuit $012'3'\ldots n'$ *represents the polynomial*
$$\frac{F(z)-F(x)}{z-x},$$
or, in the case where x is a root of the equation $F(z)=0$, *the polynomial* $F(z):(z-x)$.

92. The similar triangles considered above give
$$\frac{01}{0A_1} = \frac{A_1 2}{A_1A_2} = \frac{A_2 3}{A_2A_3} = \frac{A_3 4}{A_3A_4} = \ldots = \frac{A_{n-1}.n}{A_{n-1}A_n};$$
so that our equation may also be written,
$$0A_1 \cdot \frac{B_nA_n}{B_1A_1} = 0A_1y^{n-1} + A_1A_2\,y^{n-2} + A_2A_3\,.\,y^{n-3} + \ldots + A_{n-1}A_n.$$
This result may be interpreted as follows (Fig. 77):

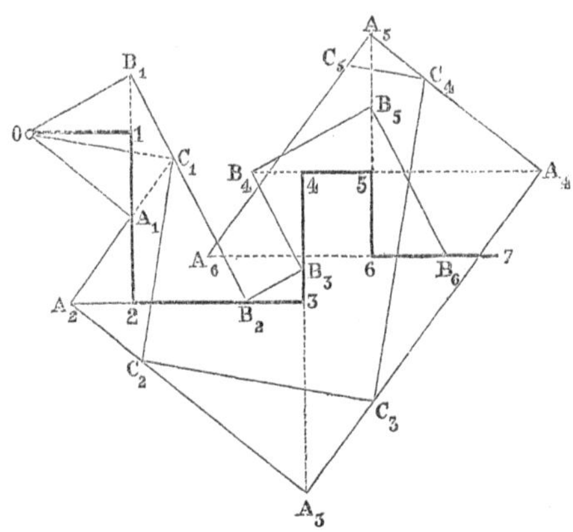

Fig. 77.

If we inscribe in the rectangular circuit $0123\ldots n+1$ of $n+1$

sides (Fig. 77, *where* $n = 6$) *two new rectangular circuits of* n *sides* $0A_1A_2\dots A_n$, $0B_1B_2\dots B_n$, *and then inscribe in the first of these a rectangular circuit of* $n-1$ *sides* $0C_1C_2\dots C_{n-1}$,

where
$$\frac{C_1A_1}{0A_1} = y = \frac{B_1 1}{01},$$

then
$$\frac{B_nA_n}{B_1A_1} = \frac{C_{n-1}A_n}{0A_1};$$

that is to say, $C_{n-1}A_n$ *is also equal to the quotient*
$$\frac{F(y)-F(x)}{y-x},$$

multiplied however by
$$\frac{0A_1}{01}.$$

In other words:

The circuit $OA_1A_2\dots A_n$ *represents the polynomial*
$$\frac{F(z)-F(x)}{z-x};$$
or, in the case where x *is a root of the equation* $F(z) = 0$, *the polynomial* $F(z):z-x$, *provided the lengths be reduced in the ratio* $OA_1:O1$.

Every rectangular circuit of n sides inscribed in the given circuit, and having the same extremities, is therefore a *resolvent* circuit in regard to the given one, because it represents the quotient, obtained by the division of the Polynomial represented by the given circuit, by one of its linear factors.

93. Again, let the entire polynomial of the nth degree $F(z)$ be represented by the circuit $0123\dots n+1$. In it let the two circuits $OA_1A_2\dots A_n$, $OB_1B_2\dots B_n$ (Fig. 78) be inscribed. We assume that both the points A_n, B_n coincide with the extremity $n+1$ of the given circuit; i.e. let $OA_1A_2\dots A_n$, $OB_1B_2\dots B_n$ be two resolvent circuits of the given circuit. Moreover let $L_1, L_2, \dots, L_{n-2}$ be the points of intersection of the pairs of sides A_1A_2, B_1B_2; A_2A_3, B_2B_3; ...; $A_{n-2}A_{n-1}$, $B_{n-2}B_{n-1}$. Then the triangles $0A_1B_1$, $L_1A_2B_2$ are similar, since their corresponding sides are perpendicular to one another; for the same reason the triangles $A_1B_1L_1$, $A_2B_2L_2$ are similar, and therefore also the quadrilaterals $0A_1B_1L_1$, $L_1A_2B_2L_2$ are similar, whence it follows that the sides $0L_1$, L_1L_2 are perpendicular to one another. In the same way we show that the angles $L_1L_2L_3$, $L_2L_3L_4, \ldots, L_{n-3}L_{n-2}\overline{n+1}$ are right angles.

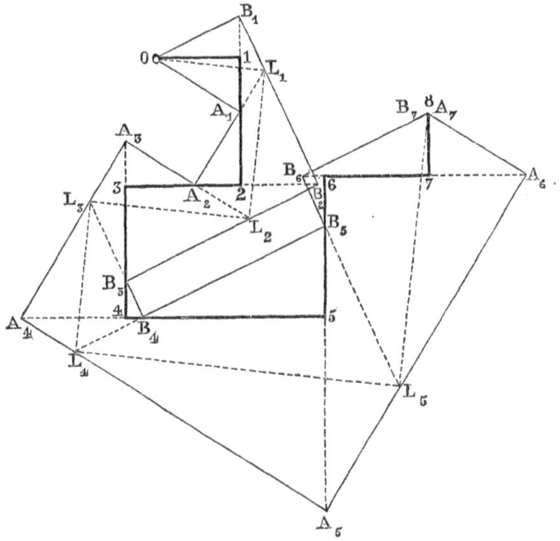

Fig. 78.

Hence it follows that the points $0L_1L_2 \ldots L_{n-2}\overline{n+1}$ are the vertices of a circuit of $n-1$ sides, which is right-angled, and is inscribed both in the circuit $0A_1A_2, \ldots$, and in $0B_1B_2 \ldots$; that is to say, $0L_1L_2 \ldots$ is a resolvent circuit in regard to each of the circuits $0A_1A_2 \ldots$, $0B_1B_2 \ldots$. In other words, if we reduce the lengths in the ratio $\frac{0L_1}{01}$, the circuit $0L_1\, L_2 \ldots L_{n-2}$ represents the polynomial of the $(n-2)^{\text{th}}$ degree

$$\frac{F(z)}{(z-x)\,(z-y)},$$

where $x = 0A_1 : 01$, $y = 0B_1 : 01$.

94. Let an equation of the second degree be given

$$a_0x^2 + a_1x + a_2 = 0.$$

After constructing the circuit 0123 (Fig. 79), whose sides 01, 12, 23 represent the coefficients a_0, a_1, a_2, it is sufficient, in order to find a root, to construct a right angle, with its vertex A upon 12, and its legs passing through 03. We describe therefore a semicircle

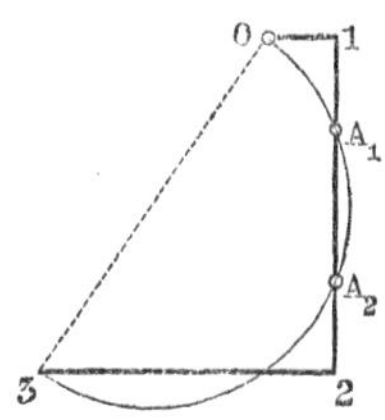

Fig. 79.

on 03 as diameter; if this cuts 12 in the points A_1, A_2, the roots of the given equation will be

$$\frac{A_1 1}{a_0}, \quad \frac{A_2 1}{a_0}.$$

From known properties of the circle we have:

$$A_2 1 = 2 A_1,$$

and therefore

$$A_1 1 + A_2 1 = 2 A_1 + A_1 1 = 21,$$

or

$$\frac{A_1 1 + A_2 1}{a_0} = -\frac{a_1}{a_0},$$

that is to say, the sum of the roots is $-\frac{a_1}{a_0}$.

Further the similar triangles $01 A_1$, $A_1 23$ give

$$01 : A_1 1 = A_1 2 : 32,$$

or

$$a_0 : A_1 1 = -A_2 1 : -a_2,$$

and therefore

$$\frac{A_1 1 \,.\, A_2 1}{a_0{}^2} = \frac{a_2}{a_0},$$

i.e. the product of the roots is equal to $\frac{a_2}{a_0}$.

A simple apparatus depending on the foregoing theorem (Art. 91) has been designed by Lill, with the object of determining the roots of a given numerical equation. The apparatus consists of a perfectly plane circular disc, which may be made of wood; upon it is pasted a piece of paper ruled in squares. In the centre of the disc, which should remain fixed, stands a pin, around which as a spindle another disc of ground glass of equal diameter can turn. Since the glass is transparent, we can, with the help of the ruled paper underneath, immediately draw upon it the circuit corresponding to the given equation. If we now turn the glass plate, the ruled paper assists the eye in finding the circuit which determines a root. A division upon the circumference of the ruled disc enables us, by means of the deviation of the first side of the first circuit from the first side of the second, to immediately determine the magnitude of the root. For this purpose the first side of the circuit corresponding to the equation must be directed to the zero point of the graduation.

CHAPTER VII.

REDUCTION OF PLANE FIGURES*.

95. To reduce a given figure to a given base b, we must transform the figure into a rectangle whose base is b, or determine a straight line f, which when multiplied by b gives the area of the given figure. Instead of constructing a rectangle on the base b we may construct a triangle on the base $2b$; the height of this triangle is the straight line f. The segment b is called the *base of reduction.*

When several figures are reduced to the same base b, their areas are proportional to the corresponding straight lines f_1, f_2, f_3, &c.; whence it follows, that the reduction of a figure to a given base is the same thing as finding its area.

Let the given figure be the triangle OAB (Fig. 80), whose base OA is denoted by a, and its height by h. Then, since the area must remain unaltered by the transformation, $fb = \frac{1}{2}ah$,

and therefore $$f = a \cdot \frac{h}{2b} = h \cdot \frac{a}{2b},$$

that is to say, we have either to multiply a by the ratio $h : 2b$, or h by the ratio $a : 2b$.

We therefore take $OC = 2b$, join C to B, and draw AD parallel to CB.

Or else, take on OB the point D, whose distance from $OA = 2b$, join DA, and draw BC parallel to DA.

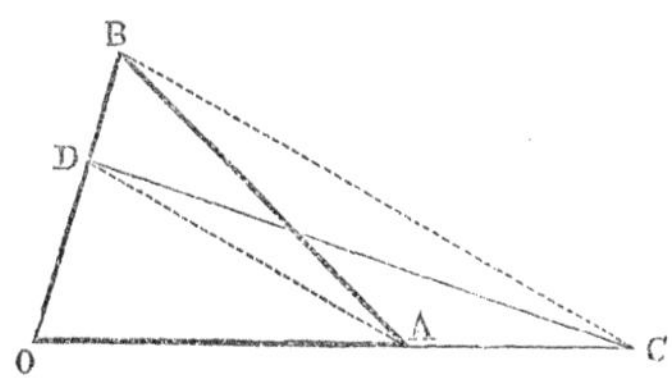

Fig. 80.

If we join CD, the triangles OAB, and OCD are equivalent, because we obtain them, if we subtract the equal triangles ADB, ADC from, or add them to the same triangle OAD (according as OC is smaller than, or greater than OA). The required segment f is therefore in the first construction the

* CULMANN, l. c., No. 15 et seqq.

height of the point D above OC, and in the second is the length OC.

96. It is not necessary that one of the dimensions f, or $2b$, should fall along a side of the given triangle. We may take as the doubled base $2b$ a straight line BC (Fig. 81) drawn from the vertex B to the opposite side OA, provided $2b$ is not less than the distance of B from OA; the corresponding height f will then be OD, the antiprojection of OA on BC*. Or if $2b$ is not greater than OA, we can take as the doubled base $2b$ a chord OD of the semicircle drawn on OA as diameter In this case the parallel BC to the supplementary chord DA is the required height f.

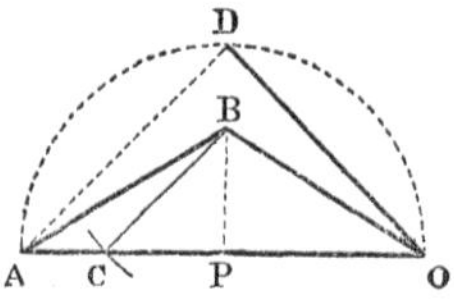

Fig. 81.

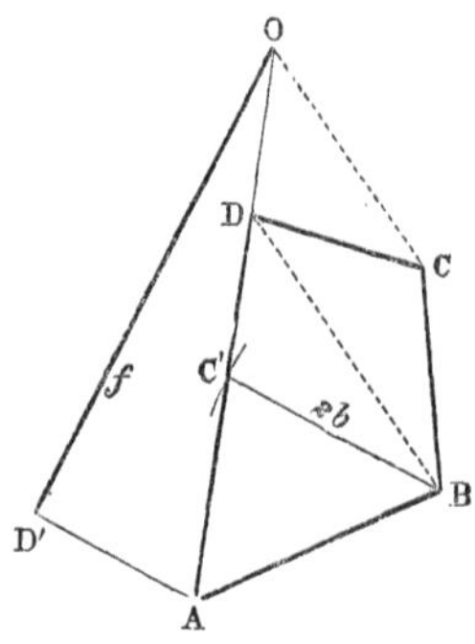

Fig. 82.

97. Let it be required to reduce the quadrilateral $ABCD$ to the base b (Fig. 82). Draw CO parallel to the diagonal BD; then the quadrilateral reduces to the triangle OAB, and we proceed as above, viz. we make $BC' = 2b$, and the antiprojection OD' of OA upon BC' is the required length f.

98. The reduction may also be performed without first reducing the given quadrilateral $ABCO$ to a triangle. Take the diagonal OB (Figs. 83, 84, 85, 86), which must not be less than $2b$, as hypothenuse, and construct the right-angled triangle ODB of which the side $BD = 2b$. Let the points A and C be projected, by means of rays parallel to OB, into A', C' on OD, the other side of the triangle OBD; the

* The triangles BPC, DOA are similar, hence $BP : BC = OD : OA$, or $h : 2b = f : a$; therefore $f = a \cdot \frac{h}{2b}$. Q. E. D.

triangles OCB, OBA are equivalent to the two triangles $OC'B$, and OBA'; but in each of these the distance of the base OC', or

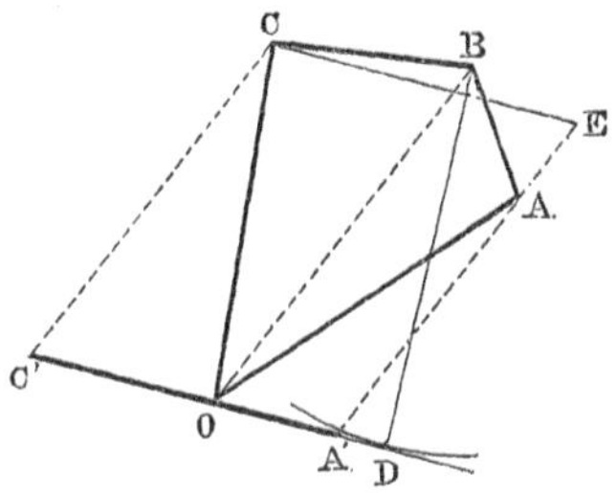

Fig. 83.

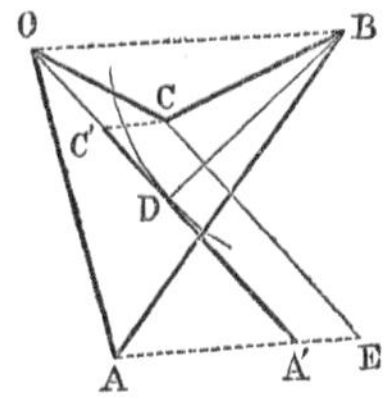

Fig. 84.

OA', from the opposite vertex $= 2b$, and therefore the required height f for the quadrilateral is equal to $OC' + A'O = A'C'$.

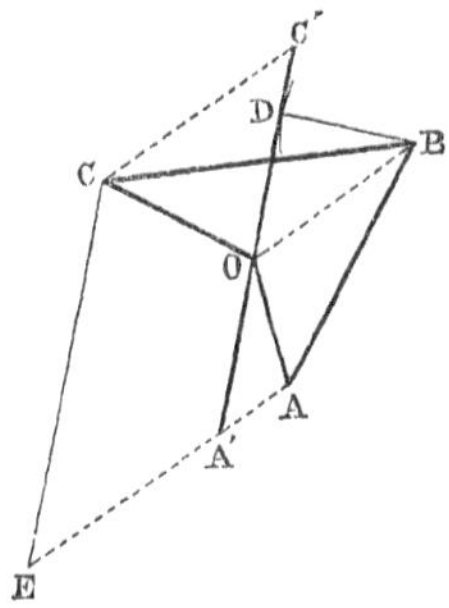

Fig. 85.

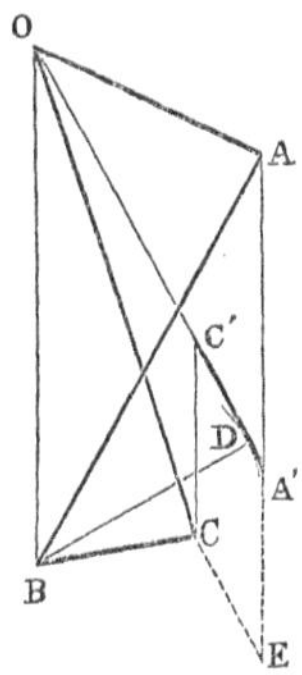

Fig. 86.

In the crossed quadrilateral (Fig. 87) if AC is parallel to BO, the points A' and C' coincide, and therefore $f = 0$. In fact, in this case the area $ABCO$ is equal to the sum of the two triangles UAB, and UCO, which are of equal area but of opposite sign.

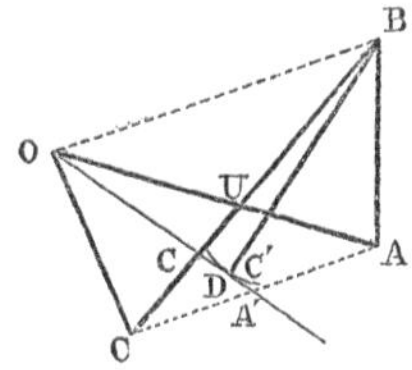

Fig. 87.

99. The length f is also equal to that segment of the straight line drawn through A or C parallel to $A'C'$ which is intercepted by the straight lines AA', CC'.

100. The foregoing construction assumes that $2b$ is not greater than the greatest diagonal OB of the quadrilateral. If $2b$ is $> OB$, the lengths $2b$ and f can be interchanged. We draw, namely, AE parallel to OB and make $CE = 2b$;

then, construct on the hypothenuse OB a right-angled triangle ODB, of which the side OD is parallel to CE; and then the other side $BD = f$.

101. In order to reduce a polygon to a given base, whether its periphery is self-cutting or not, we begin by reducing it to an equivalent quadrilateral. We then apply the above construction to the quadrilateral and thus obtain the segment f, which multiplied by the base b gives the area of the polygon proposed.

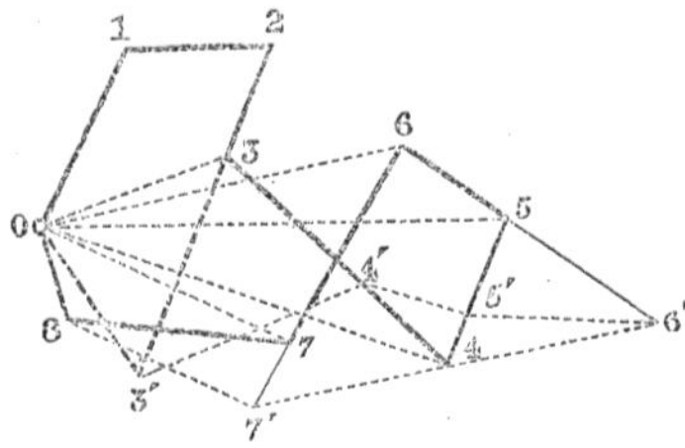

Fig. 88.

Let the given polygon be 0123456780 (Fig. 88). Draw the straight line 8 7′ parallel to the diagonal 07,

,,	,,	7′6′	,,	,, 06,
,,	,,	6′5′	,,	,, 05,
,,	,,	5′4′	,,	,, 04,
,,	,,	4′3′	,,	,, 03,

and the polygon is successively transformed into the equivalent polygons 01234567′, 0123456′, 012345′, 01234′, 0123′, each of them having a side less than the preceding one*. We finally arrive at the quadrilateral 0123′.

102. In this construction the new sides 07′, 06′, 05′,... of the reduced polygons are rays proceeding from the fixed vertex 0. But we can also proceed in such a manner, that all the new vertices 7′, 6′, 5′, &c. lie on a specified side. If we have, for example, the polygon $Aabcde\ C012345$, and if we draw

11′	parallel to	20	till each intersects the side $0C$,
22′	,,	31′	
33′	,,	42′	
44′	,,	53′	
$5D$	,,	$A4'$	

* The triangles 087, 077′ are equivalent, because the straight lines 07, 87′ are parallel; if we take the first triangle away from the given polygon, and add the second one to it, we obtain the new polygon 01234567′0. And so on.

we determine a straight line *AD* which can be substituted for the crooked line *A*543210.

For, since 11′, 20 are parallel, the triangles 120 1′20 are equivalent, and if we subtract the former from the given polygon, and add the latter to it, the polygon reduces to *AabcdeC* 1′2345. Similarly, from the equivalence of the triangles 1′23, 1′2′3 the latter polygon reduces to *Aabcde C* 2′345, and proceeding in this manner we finally arrive at the polygon *Aabcde CD*.

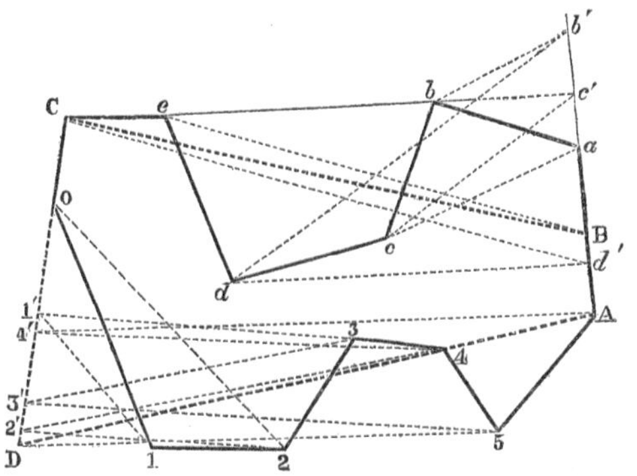

Fig. 89.

In order to effect a like transformation for the crooked line *AabcdeC*, we draw

bb′ parallel to *ca*
cc′ „ *db′*
dd′ „ *ec′*
eB „ *Cd′*
} till each intersects the side *Aa*,

and now the whole polygon *Aabcde C*012345 is reduced to the equivalent quadrilateral *ABCD*.

103. This is the easiest and most convenient way of finding the areas of figures, the perimeters of which take the most different forms. With a little practice we learn to perform the reduction quite mechanically, and without paying any attention to the actual form of the proposed circuit. This construction moreover permits us to take account of signs, so that in dealing with areas of different signs, the result gives the actual sign belonging to their sum without further trouble*. Take, for example, the self-cutting circuit (Fig. 90) $ABC\,01234$, which represents the cross section of an embankment and excavation in earthwork.

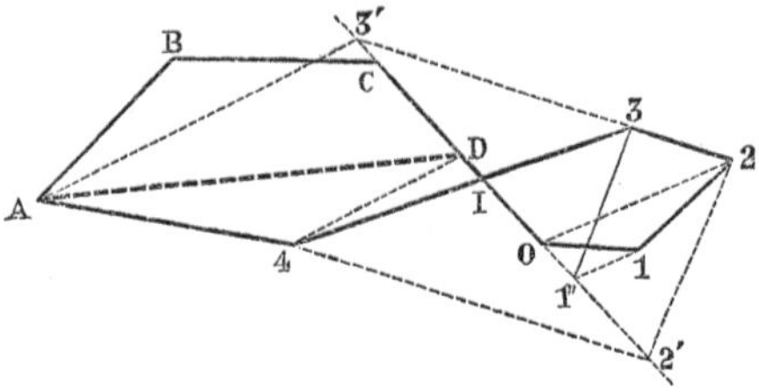

Fig. 90.

Draw	11′	parallel to	20 ,
	22′	"	31′,
	33′	"	42′,
	$4D$	"	$A3'$,

until they meet the side $C0$, then the given polygon is reduced to the equivalent quadrilateral $ABCD$, which therefore represents the difference between the areas $ABCI4$ of the embankment and $I\,0123$ of the excavation, which have necessarily different signs. The circuit $ABCD$ has the same sign as the circuit $ABCI4$, or as the circuit $I\,0123$, according as the embankment or the excavation is the larger.

104. *Circular Figures.* A sector of a circle (Fig. 91) OAB is equivalent to a triangle OAC, with its vertex at the centre O and its base a portion AC of the tangent equal to the arc AB. In order to obtain approximately the length of the arc AB measured along the tangent, we take an arc a, so small that it can without any sensible error be replaced by its chord a; we then apply the chord a to the given arc AB starting from its extremity B, and continue doing so as long as necessary till we reach A or a point A' very near A. Then starting from A or A' we set off the chord a the same number of times along the tangent AC†. The sector OAB is now replaced by the rectilinear triangle OAC.

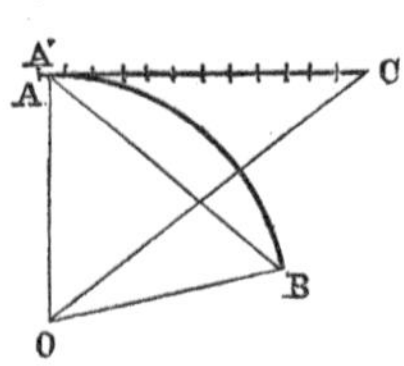

Fig. 91.

* Culmann, l. c., No. 17.

† Culmann, l. c., No. 21. In Chapter IX is given a method of Rankine for the approximate rectification of circular arcs, and also some methods of Professor Sayno.

The segment AB (i.e. the area between the arc AB and its chord) is the difference of the two triangles OAC, OAB, and is equivalent therefore to the crossed quadrilateral $OBAC$.

105. It is not necessary that the tangent upon which we set off the arc should pass through an extremity of the arc; instead of doing so it may (Fig. 92) touch the arc at any other

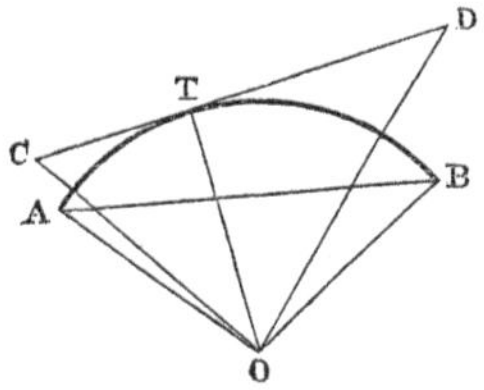

Fig. 92.

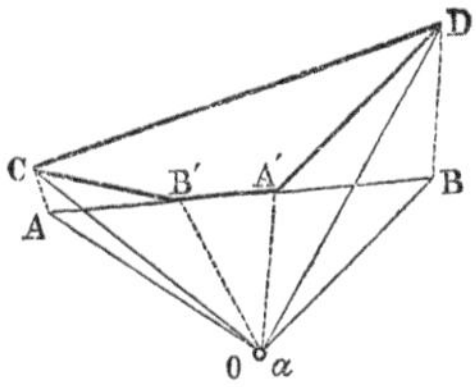

Fig. 93.

point T. In such case we set off the arc AT on CT, and BT on DT. The sector OAB is transformed into the triangle OCD, and the segment AB is the difference between OCD and OAB, i.e. is equal to the doubly crossed figure $OCDOBAO$, which may be considered as a hexagon (Fig. 93) with two coincident vertices at O. If we draw OB' and OA' respectively parallel to AC and BD, the triangles OAC, OBD are transformed into the two other triangles $B'AC$, $A'BD$, and therefore the segment is equal to the quadrilateral $A'B'CD$.

106. *Example—*

Let the figure to be reduced be the four-sided figure $ACD\,3$, contained between two non-concentric circular arcs, AC and $3\,D$, and the straight lines CD, $A\,3$ (Fig. 94).

Let 0, 1 be the centres of the two circles; the given figure is then equal to the sector $0\,AC$—the sector $13\,D$—the quadrilateral $0A1C$. Change the sectors into the triangles $0AB$, 132, by setting off the two arcs along their respective tangents AB and 32, starting from corresponding extremities A, 3; and now the given figure is

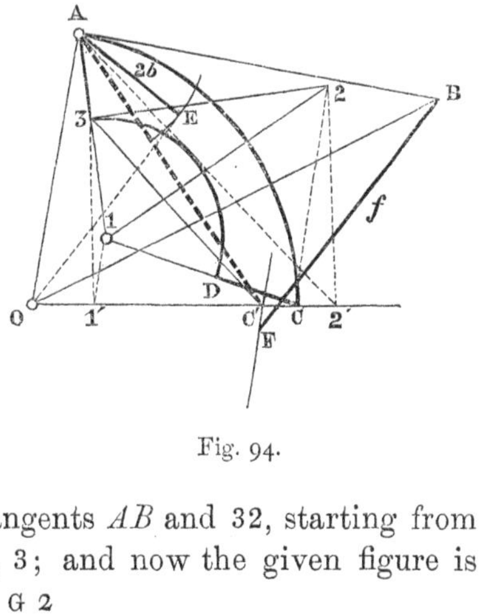

Fig. 94.

equal to the triangle $0AB$—the area $0A321C0$, i.e. is equal to the self-cutting polygon $AB0C123A$.

$$\left.\begin{array}{lll} \text{Draw } 11' & \text{parallel to} & 2C \\ \quad\;\; 22' & \quad ,, & 31' \\ \quad\;\; 3C' & \quad ,, & A2' \end{array}\right\} \text{till they cut the fixed side } C0,$$

and the polygon reduces to the crossed quadrilateral $AB0C'$. The area bf of this quadrilateral is found in the usual manner; i.e. on the diagonal $A0$ as hypothenuse a right-angled triangle is constructed, of which one side $AE = 2b$; the length f is then the distance, measured parallel to the other side, of the point B from a straight line parallel to $A0$ and passing through C'.

107. As another example suppose we wish to determine the area of Fig. 95, which represents the cross section of a so-called U-iron.

It consists; (1) of a lune-shaped area $AEA'F$, bounded by two circular arcs, one having U as centre, the other O; (2) of a crown-shaped piece $CBFB'C'$ bounded by two concentric circular arcs BB', CC' drawn with O as centre; (3) of two equal rectilinear pieces* $BCJIH$ and $B'C'J'I'H'$, symmetrically situated with regard to the straight line $OUFE$, which is an axis of symmetry for the whole figure.

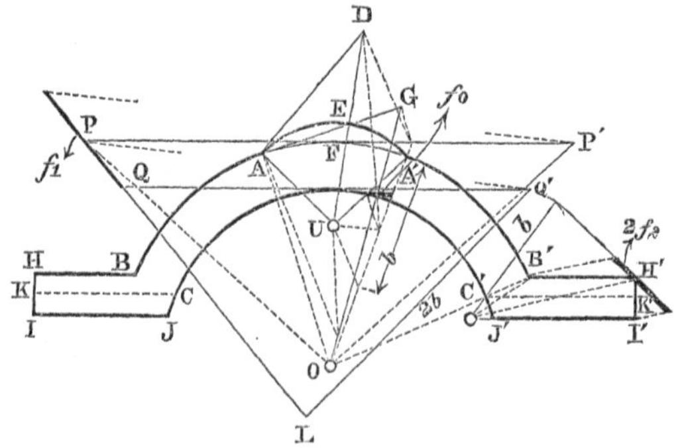

Fig. 95.

The lune is equal to the sector $UAEA'$ plus the quadrilateral $OAUA'$ minus the sector $OAFA'$, i.e. it equals the sum $UAEA' + OAUA' + AOA'F$. After transforming the two sectors spoken of into the triangles UAD, OAG (where AD, AG are the arcs AEA' and AFA' set off along their respective initial

* We say rectilinear, because we suppose the small arcs CJ, $C'J'$ to be replaced by their chords.

tangents), the lune becomes equal to the sum $UAD + OAUA'$ $+ AOG$, or finally, if we merge these three circuits in one, it is equal to the area of the circuit $ADUA'OAOGA$. Here we can neglect the part OA, which is twice passed over in opposite senses; and consequently (Art. 23) the lune is equal to the self-cutting hexagon $ADUA'OGA$.

The crown-piece we consider as the difference of the sectors OBB', OCC'. After setting off the arcs along their middle tangents PP', QQ', since PQ, $P'Q'$ both pass through O, the crown becomes equal to the trapezium $PP'Q'Q$, which is the difference between the two triangles OPP', OQQ', equivalent to the two sectors in question.

If we now reduce the hexagon $ADUA'OG$, the trapezium $PP'Q'Q$, and the two pentagons $BCJIH$ to a common base b, and find their corresponding segments to be f_0, f_1, $2f_2$, then $b(f_0 + f_1 + 2f_2)$ will be the required area of the given figure*.

Or we may consider the given figure as an aggregate of triangles and trapeziums made up in the following way

$$UAD + 2\,OAU - OAG + OPP' - OQQ' + 2\,BCKH + 2\,CJIK,$$

where CK is drawn parallel to BH, and JI. We consider the areas of these triangles and trapeziums as the products of two

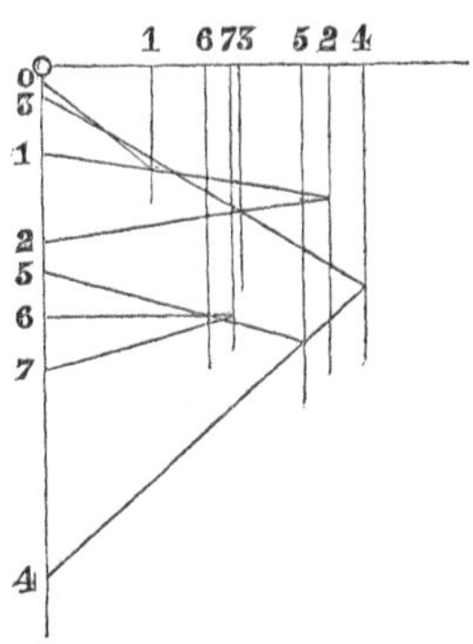

Fig. 95 *a*.

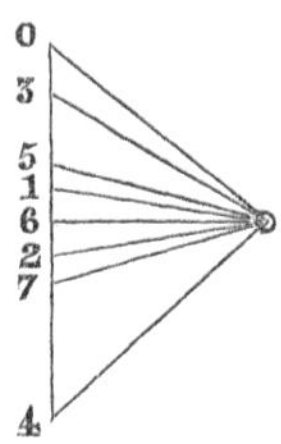

Fig. 95 *a*.

factors, and reduce these products to a base b by means of the multiplication polygon (Fig. 95 *a*). It is, of course, understood that for each area to be subtracted, one of the two factors must be taken negatively.

* In Fig. 95 we find $2f_2$ directly, if in the reduction of the figure $B'C'J'I'H'$ we substitute b for $2\,b$.

108. *Curvilinear figures in general**. It is a well-known property of the parabola, that a parabolic segment (Fig. 96) is equivalent to $\frac{4}{3}$ of the triangle, whose base is that chord of the parabola which forms the base line of the segment, and the vertex of which is that point of the arc where the tangent is parallel to the base; that is to say, the segment of a parabola is equal to a triangle whose base is the chord, and whose altitude is $\frac{4}{3}$ the *Sagitta*: where we understand by Sagitta the perpendicular distance between the chord and that tangent of the arc which is parallel to the chord.

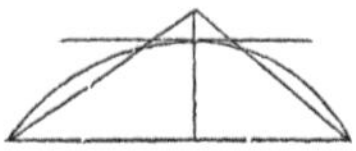

Fig. 96.

109. One method then of reducing curvilinear figures, consists in considering each small portion of the curved periphery to be a parabolic arc.

If a curved line (Fig. 97) is divided into small arcs each of which may be approximately regarded as a parabolic arc, and if the parabolic segments between these arcs and their respective chords are reduced to triangles on these chords as bases; then the vertices of all these triangles can be taken anywhere at pleasure on the straight lines drawn parallel to the chords at distances from them equal to $\frac{4}{3}$ their respective Sagittae. Let these vertices be taken so that the vertex of each new triangle lies on the prolongation of one side of the preceding triangle, i.e. so that the vertices of two successive triangles and the point of intersection of their bases always lie in the same straight line. Then the curvilinear circuit is reduced to an equivalent rectilinear circuit formed of sides whose number is equal to that of the parabolic segments into which the given circuit was divided. The rectilinear circuit or polygon is next reduced to its equivalent quadrilateral, and finally this is reduced to the given base in the manner previously explained.

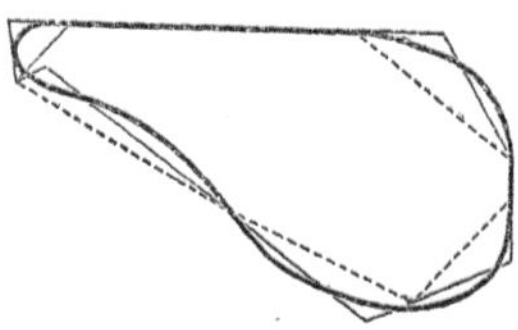

Fig. 97.

110. Suppose, for example, we wish to replace the irregular boundary line AB between two fields by another consisting of two rectilinear segments which form an angle with its extremities at A and B (Fig. 98). We consider the curve AB

* Culmann, l. c., No. 23.

and the straight line BA as a circuit, and reduce it to a triangle on the base BA.

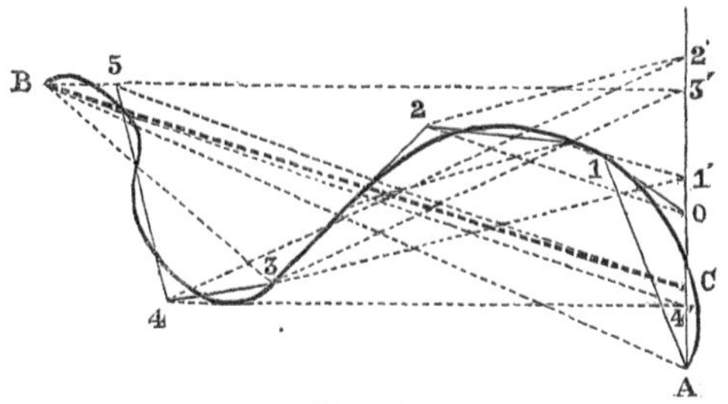

Fig. 98.

For this purpose we divide the curve into small arcs; draw their chords and for the segments thus formed substitute triangles, by the method we have just given above. In this way we transform the given circuit into the rectilinear polygon $A012345B$. Then we draw

$11'$ parallel to	20		
$22'$ "	$31'$		
$33'$ "	$42'$	till each cuts the fixed line $A0$;	
$44'$ "	$53'$		
$5C$ "	$B4'$		

and thus transform the polygon into the triangle ACB. We have therefore substituted the two rectilinear segments AC, CB for the given irregular line. The point C can be displaced at pleasure along a line parallel to AB, since by doing so we do not alter the area ABC.

111. The reduction of areas to a given base furnishes another construction for the resultant of a number of segments A_1B_1, A_2B_2, &c., &c. given in magnitude, sense, and position (Fig. 99). Take a point O as the initial point of a polygonal circuit whose sides are respectively equipollent to the given segments; let N be its final point. Now transform the triangles OA_1B_1, OA_2B_2, &c., by reducing them to a common base ON, and let them be so transformed that they have a vertex at O, and the side opposite to it equipollent to ON. Then the sum,

$$OA_1B_1 + OA_2B_2 + \&c., \&c.,$$

will also have been transformed into a triangle OAB, where AB is equipollent to ON. The segment AB is the required resultant (Art. 46).

In order to effect the above-mentioned transformation, it will be convenient to take the initial points A_1, A_2, &c., &c. of

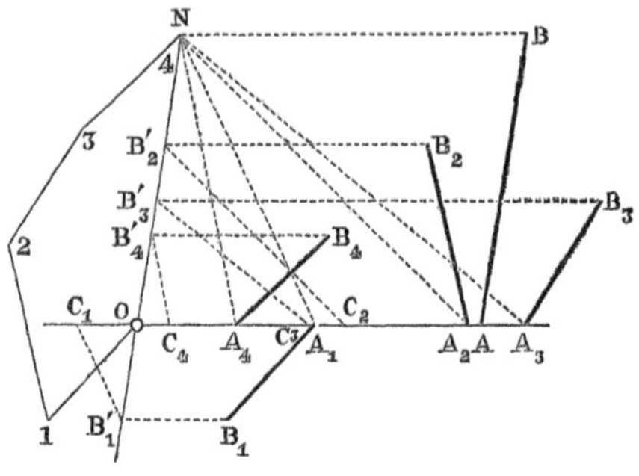

Fig. 99.

the segments in a line with O. Project the points B_1, B_2, &c., &c. into the points B_1', B_2', &c., &c. upon ON, by rays parallel to $OA_1A_2 \ldots$, and then the triangles OA_1B_1, OA_2B_2, &c., are transformed into the triangles OA_1B_1', OA_2B_2', &c. Then draw the straight lines $B_1'C_1, B_2'C_2$, &c., parallel to NA_1, NA_2, &c., respectively, and let the points in which they cut the straight line $OA_1A_2 \ldots$ be C_1, C_2, &c.

We thus obtain the triangles OC_1N, OC_2N, &c., respectively equivalent to OA_1B_1', OA_2B_2', &c., &c. Therefore, if the segment $OA = OC_1 + OC_2 +$ &c., is taken on $OA_1A_2 \ldots$, and if through A the straight line AB is drawn equipollent to ON, then OAB is equal to $OA_1B_1 + OA_2B_2 +$ &c.

CHAPTER VIII.

CENTROIDS.

112. Let us suppose that, in the theorems of Articles 43 and 44, all the points $B_1, B_2, \dots, B_n$ coincide in a single point G; these theorems may then be stated as follows:

If $A_1 G$, $A_2 G$, $A_3 G, \dots, A_n G$ are n segments, whose resultant vanishes, and O is any arbitrarily assumed point in the plane, the resultant of the segments OA_1, $OA_2, \dots, OA_n$ is equal (equipollent) to n times the segment OG (Fig. 100).

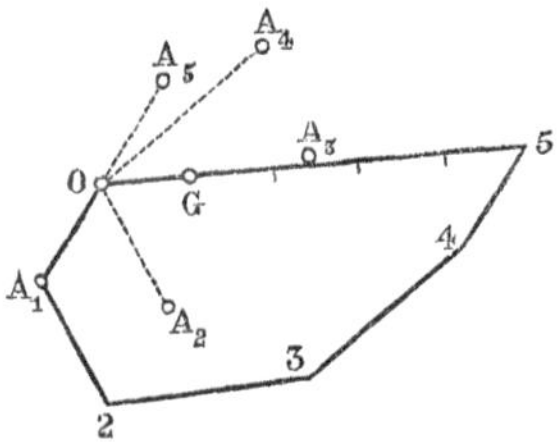

Fig. 100.

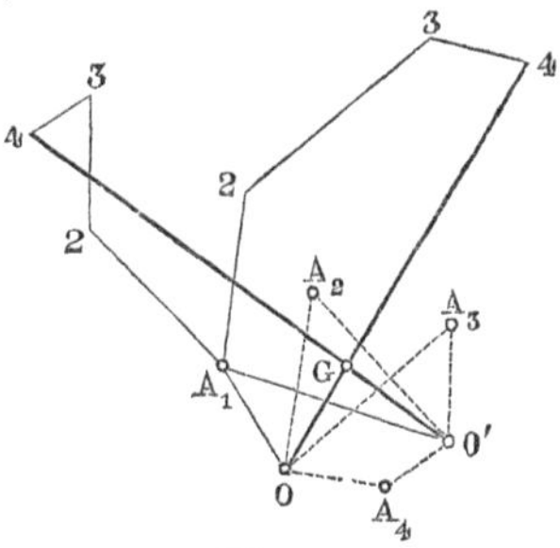

Fig. 101.

Conversely:

*Let there be given n points $A_1, A_2, \dots, A_n$, and let the resultant of the straight lines OA_1, $OA_2, \dots, OA_n$, which join the pole O to the given points, be equal to n times the straight line OG drawn from O to G, then the same equality holds for any other pole O'; that is to say, the resultant of the straight lines $O'A_1$, $O'A_2, \dots$, $O'A_n$ is equal to n times the segment $O'G$; and the resultant of the straight lines GA_1, $GA_2, \dots, GA_n$ is equal to zero**.

113. The point G is called the *Centroid* of the points A_1, $A_2, \dots, A_n$. Let the (Fig. 101, where $n = 4$) points $A_1, A_2, \dots$, A_n be given, to construct their centroid G we proceed as follows. An arbitrary pole O is taken, and a circuit

* Grassmann, l. c., p. 141. Chelini, *Sui centri de' sistemi geometrici* Raccolta scientifica; Roma, marzo 1849), § 1.

$OA_1 23 \ldots n$ constructed, whose initial point is O and whose successive sides are equipollent to the segments OA_1, OA_2, ..., OA_n. The straight line On, which closes the circuit, passes through the point G, and $OG = \frac{On}{n}$. Instead of dividing On into n equal parts in order to obtain G, we may construct a second circuit starting from another initial point O'; the straight line which closes this new circuit will cut On in the required point G.

114. The system of n given points cannot have another centroid G'. For if both the resultant of GA_1, GA_2, ..., GA_n, and the resultant of $G'A_1$, $G'A_2$, ..., $G'A_n$ should vanish, then the general resultant of all the segments GA_1, A_1G', GA_2, A_2G', ..., GA_n, A_nG' would also vanish. But if we combine the two segments GA_r, A_rG', we obtain the segment GG'; and therefore GG' must vanish, that is to say, G' must coincide with G.

115. Again, if in the proposition of Article 45, all the points B_1, B_2, ..., B_n are supposed to coincide with a single point G, the theorem may be stated as follows:

If G is the centroid of the points 1, 2, 3, ..., n, *and if all these points are projected by means of parallel rays into the points* G', 1′, 2′, 3′, ..., n' *upon one straight line, then the sum of the straight lines* 11′, 22′, 33′, ..., nn' *is equal to n times the straight line GG'* (Fig. 102).

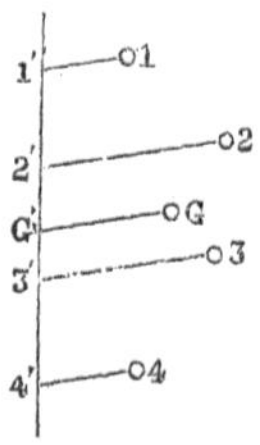

Fig. 102.

As a result of this proposition, since rr' is the (oblique) distance of the point r from the straight line upon which we project, the point G is also called the *centre of mean distances** of the given points 1, 2. 3, ..., n.

116. Instead of supposing in the propositions of Articles 43 and 44, that all the points B_1, B_2, ..., B_n coincide with a single point G, we now imagine some of them B_1, B_2, ..., B_i to remain distinct, and the rest to coincide with a single point G; so that the resultant of the segments A_1B_1, A_2B_2, ..., A_iB_i, $A_{i+1}G$, ..., A_nG vanishes; and, whatever the position of O may be, the resultant of OA_1, OA_2, ..., OA_n is equal to the resultant of OB_1, OB_2, ..., OB_i, $(n-i) . OG$. The

* Carnot, *Corrélation des figures de géométrie* (Paris, 1801), No. 209.

first of these equalities does not change, if we substitute for the segment $A_r B_r$ the two others $A_r G$, GB_r or $A_r G, -B_r G$; the second equality will also continue to subsist if we add to both resultants the segments $B_1 O, B_2 O, \ldots, B_i O$, so that it becomes an equality between the resultant of $OA_1, OA_2, \ldots, OA_n, B_1 O, B_2 O, \ldots, B_i O$ and the resultant of $OB_1, OB_2, \ldots, OB_i, B_1 O, B_2 O, \ldots, B_i O, (n-i).OG$; that is, between the resultant of $OA_1, OA_2, \ldots, OA_n, -OB_1, -OB_2, \ldots, -OB_i$ and $(n-i) . OG$.

Hence:

If the resultant of the segments $A_1 G, A_2 G, \ldots, A_n G, -B_1 G, -B_2 G, \ldots, -B_i G$ vanishes; then, for any point O whatever, the resultant of the segments $OA_1, OA_2, \ldots, OA_n, -OB_1, -OB_2, \ldots, -OB_i$ is equal to $(n-i) . OG$: and conversely if this equality subsists for any pole O, it will also hold for any other pole O', and the resultant of the segments $A_1 G, A_2 G, \ldots, A_n G, -B_1 G, -B_2 G, \ldots, -B_i G$ will vanish.

117. Now let us assume that of the n points $A_1, A_2, \ldots, A_n$ some coincide with one point, others (in like manner) with a second point, and so on; and that the points $B_1, B_2, \ldots, B_i$ also unite in groups and coincide. Then if we use $a_1, a_2, a_3, \ldots$ to denote positive or negative integral numbers whose sum is m, the foregoing proposition may be stated as follows:

If the points $A_1, A_2, A_3, \ldots$ and the point G are so situated, that the resultant of the segments $a_1 . A_1 G, a_2 . A_2 G, a_3 . A_3 G, \ldots$ vanishes, then, wherever the pole O may be, $m . OG$ will be equal to the resultant of the segments $a_1 . OA_1, a_2 . OA_2, a_3 . OA_3, \ldots$ &c.

And conversely:

If this property holds for any pole O, viz. that $m . OG$ is equal to the resultant of $a_1 . OA_1, a_2 . OA_2, a_3 . OA_3, \ldots$, then the same property holds for every other pole O', that is to say, the resultant of $a_1 . O'A_1$ $a_2 . O'A_2, a_3 . O'A_3, \ldots$ is equal to $m . O'G$; and the resultant of the segments $a_1 . GA_1, a_2 . GA_2, a_3 . GA_3, \ldots$ vanishes.

118. The point G is called the centroid *of the points $A_1, A_2, A_3, \ldots$ weighted with the coefficients $a_1, a_2, a_3, \ldots$.* For shortness however we say that G is the centroid *of the points $a_1 . A_1$, $a_2 . A_2, a_3 . A_3, \ldots$*, writing before each point the coefficient which belongs to it.

119. Furthermore, from the proposition of Article 45 we obtain the following theorem:

If G is the centroid of the points $a_1 . A_1, a_2 . A_2, a_3 . A_3, \ldots$ and if, *by means of parallel rays, the points G, A_1, A_2, A_3, ... are projected into the points G', A'_1, A'_2, A'_3, ... which lie on a straight line, then the sum of the straight lines* $a_1 . A_1A_1'$, $a_2 . A_2A_2'$, $a_3' . A_3A_3'$, ... *is equal to* $m . GG'$, *where* $m = a_1 + a_2 + a_3 + \ldots$.

On account of this property G is also called the *centre of mean distances of the points* $a_1 . A_1$, $a_2 . A_2$, $a_3 . A_3$, ...*.

120. Hitherto the coefficients a_1, a_2, a_3, ... have been positive or negative *integral numbers*; we shall now extend the idea of a *Centroid* to the case where a_1, a_2, a_3, ... are *any numbers whatever*, or rather parallel segments proportional to any given homogeneous magnitudes

Let then the points A_1, A_2, A_3, ... be given, weighted with the numbers or parallel segments a_1, a_2, a_3, ... Project the given points on to a straight line p', by means of rays parallel to some arbitrarily chosen direction, into A'_1, A'_2, ...; and by means of rays parallel to another direction chosen at pleasure, project the same points into A''_1, A''_2, A''_3, ... on a second straight line p'', not parallel to p'. Now determine a straight line r' parallel to p', such that the distance from r' to p' measured parallel to the rays $A_1A'_1$, $A_2A'_2$, A_3A_3', ..., is equal to

$$\frac{a_1 . A_1A'_1 + a_2 . A_2A'_2 + a_3 . A_3A'_3 + \ldots}{a_1 + a_2 + a_3 \ldots};$$

similarly determine a straight line r'' parallel to p'', such that the distance from r'' to p'', measured parallel to the rays $A_1A''_1$, A_2A_2'', ..., is equal to

$$\frac{a_1 . A_1A''_1 + a_2 . A_2A''_2 + a_3 . A_3A''_3 + \ldots}{a_1 + a_2 + a_3 + \ldots}.$$

Let G denote the point of intersection of the straight lines r', r'', and G', G'' the projections of G upon the straight lines p', p'' (by means of rays parallel to AA', AA'' respectively), then we shall have:

$$a_1 . A_1A_1' + a_2 . A_2A'_2 + a_3 . A_3A'_3 + \ldots = (a_1 + a_2 + a_3 + \ldots) . GG'$$
$$a_1 . A_1A''_1 + a_2 . A_2A''_2 + a_3 . A_3A''_3 + \ldots = (a_1 + a_2 + a_3 + \ldots) . GG''.$$

Next, let p''' be any third given line, let us project upon it the given points and the point G, into the points A'''_1, A'''_2, A'''_3,..., G''', by rays parallel to a new direction. Between the three rays which project the same point A_1 or A_2 or A_3, there

* L'HUILIER, *Élémens d'analyse géométrique et d'analyse algébrique etc.* (Paris, 1809), § 2.

exists (Article 16) a linear relation with constant coefficients, i.e. we have:

$$k'.A_1A'_1+k''.A_1A''_1+k'''.A_1A'''_1=k,$$
$$k'.A_2A'_2+k''.A_2A''_2+k'''.A_2A'''_2=k,$$
$$k'.A_3A'_3+k''.A_3A''_3+k'''.A_3A'''_3=k,$$
$$\cdot\quad\cdot\quad\cdot\quad\cdot\quad\cdot\quad\cdot\quad\cdot\quad\cdot\quad\cdot\quad\cdot\quad\cdot\quad\cdot\quad\cdot$$
$$k'.GG'+k''.GG''.+k'''.GG'''=k.$$

Multiply these equations by $a_1, a_2, a_3, \ldots, -(a_1+a_2+a_3+\ldots)$ respectively, and add the products; then we obtain, taking the equations already established into account,

$$k'''.\{a_1.A_1A'''_1+a_2.A_2A'''_2+a_3.A_3A'''_3+\ldots$$
$$-(a_1+a_2+a_3+\ldots).GG'''\}=0,$$

or,

$$a_1.A_1A'''_1+a_2.A_2A'''_2+a_3.A_3A'''_3+\ldots=(a_1+a_2+a_3+\ldots).GG'''.$$

That is to say:

If we project the points A_1, A_2, A_3, ..., G upon any straight line whatever by means of rays, which are parallel to an arbitrarily chosen direction, then the product of the ray which projects G by $(a_1+a_2+a_3+\ldots)$ is equal to the sum of the products formed by multiplying each of the rays which project $A_1, A_2, A_3\ldots$ by a_1, a_2, a_3, ... respectively.

We call the point G, so defined, the *centroid* of the points A_1, A_2, A_3,... loaded with the numbers or segments a_1, a_2, a_3, ...

The centroid does not change if we substitute for the coefficients a_1, a_2, a_3, ... others proportional to them, for by so doing we do not change the ratios of a_1, a_2, a_3..., to $a_1+a_2+a_3+\ldots$.

121. *If the points A_1, A_2, A_3, ..., and G are projected, by means of rays parallel to a straight line p'', on to another straight line p', and if we use O' to denote any point whatever of p', we have identically:*

$$a_1.O'A'_1+a_2.O'A'_2+a_3.O'A'_3+\ldots=(a_1+a_2+a_3+\ldots)O'G'.$$

If we draw through O' a straight line parallel to p'', and project on to it, by rays parallel to p', the points A_1, A_2, A_3, ..., G into the points A''_1, A''_2, A''_3, ..., G'', we have the identities

$$A_1A''_1=A'_1O',\ A_2A''_2=A'_2O',\ A_3A''_3=A'_3O'\ \ldots,\ GG''=G'O';$$

but from the foregoing theorem we have

$$a_1.A_1A''_1+a_2.A_2A''_2+a_3.A_3A''_3+\ldots=(a_1+a_2+a_3+\ldots).GG'',$$

and therefore the above proposition is true.

122. *If O is an arbitrary point, the resultant of the segments $a_1.OA_1$, $a_2.OA_2$, $a_3.OA_3$, ..., is $(a_1+a_2+a_3+\ldots).OG$.*

By the segment $a.OA$ we understand a segment parallel to

OA, drawn either in the sense of OA or in the opposite sense, according as a is positive or negative, and whose magnitude is equal to that of OA increased in the ratio of $a:1$. Draw through the point O a straight line p', and project upon it, by means of parallel rays, the points $A_1, A_2, A_3, \ldots, G$ into the points $A'_1, A'_2, A'_3, \ldots, G'$. Then the segment OA is the resultant of the segments OA', $A'A$, and therefore if we increase these segments in the ratio $a:1$, the resultant of $a.OA'$, $a.A'A$ will be $a.OA$. It follows, that the resultant of $a_1.OA_1, a_2.OA_2, a_3.OA_3, \ldots$, may be obtained by combining all the segments $a_1.OA'_1, a_2.OA'_2, a_3.OA'_3, \ldots$ with the segments $a_1.A_1'A_1$, $a_2.A'_2A_2, a_3.A'_3A_3, \ldots$. But the resultant (i.e. the sum) of $a_1.OA'_1, a_2.OA'_2, a_3.OA'_3 \ldots$ is $(a_1+a_2+a_3+\ldots).OG'$, and the resultant (or sum) of $a_1.A'_1A_1$, $a_2.A'_2A_2$, $a_3.A'_3A_3, \ldots$ is $(a_1+a_2+a_3+\ldots).G'G$; therefore the resultant of the segments $a_1.OA_1, a_2.OA_2, a_3.OA_3, \ldots$, can be obtained by combining the two segments $(a_1+a_2+a_3\ldots).OG'$, $(a_1+a_2+a_3\ldots).G'G$, and consequently, it coincides with the segment $(a_1+a_2+a_3+\ldots).OG$.

123. *If H is the centroid of the points $a_1.A_1, a_2.A_2, a_3.A_3, \ldots$, and K the centroid of the points $\beta_1.B_1, \beta_2.B_2, \ldots$, then the centroid of all the given points $a_1.A_1, a_2.A_2, \ldots, \beta_1.B_1, \beta_2.B_2, \ldots$ coincides with the centroid of the two points $m.H$, $n.K$, where*

$$m=(a_1+a_2+\ldots),\quad n=(\beta_1+\beta_2+\ldots).$$

For, taking an arbitrary pole O, if we combine the straight line $m.OH$, the resultant of the segments $a_1 OA_1, a_2 OA_2, \ldots$, with the straight line $n.OK$, the resultant of the segments $\beta_1.B_1, \beta_2.B_2, \ldots$, we find that $(m+n).OG$, the resultant of $m.OH$, and $n.OK$, is also the resultant of all the segments $a_1.OA_1, a_2.OA_2, \ldots, \beta_1.OB_1, \beta_2.OB_2, \ldots$.

124. *If all the points $A_1, A_2, A_3, \ldots$ lie on a straight line, their centroid G lies in the same straight line.*

This is clear, if we take the pole O upon the straight line $A_1A_2A_3\ldots$; for then all the segments $a_1.OA_1, a_2.OA_2, a_3.OA_3, \ldots$ lie in this straight line, and therefore also their resultant $m.OG$ lies in the same straight line.

From this it follows:

If we project $A_1, A_2, A_3, \ldots, A_n, G$ upon an arbitrarily chosen straight line into the points $A'_1, A'_2, A'_3 \ldots, A'_n, G'$, then the point G' is the centroid of the points $a_1.A'_1, a_2.A'_2, a_3.A'_3 \ldots, a_n.A'_n$.

Let there be only two points A_1, A_2 (Fig. 103, where the segments a_1, a_2 are simply denoted by the numbers 1, 2) with coefficients a_1, a_2, then their centroid G is a point of the straight line A_1A_2. Since the resultant of the straight lines $a_1 . GA_1$, $a_2 . GA_2$ is equal to zero, we have

$$a_1 . GA_1 + a_2 . GA_2 = 0,$$

or

$$A_1G : GA_2 = a_2 : a_1,$$

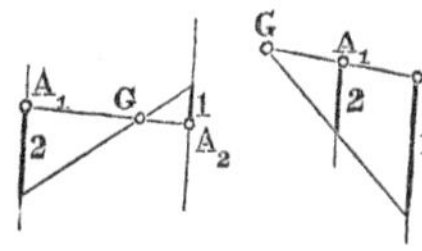

Fig. 103.

and therefore

$$A_1G : GA_2 : A_1A_2 = a_2 : a_1 : a_2 + a_1;$$

that is to say, the point G divides the segment A_1A_2 into two parts, which are inversely proportional to the numbers a_1, a_2, and it lies inside or outside the given segment, according as a_1, a_2 have the same or opposite signs.

If $a_1 = a_2$, then $A_1G = GA_2$, i.e. G is the middle point of A_1A_2. If $a_1 + a_2 = 0$, we obtain from the proportion $A_1G : A_1A_2 = a_2 : a_1 + a_2$ the value $A_1G = \infty$, i.e. G is the point at infinity of the straight line A_1A_2.

125. Let there be three given points A_1, A_2, A_3 not in one straight line (Fig. 104); and let a_1, a_2, a_3, whose sum is not zero, be their coefficients. The centroid of the points $a_2 . A_2$, $a_3 . A_3$ is a point B_1 on the straight line A_2A_3, and the centroid of the given points $a_1 . A_1$, $a_2 . A_2$, $a_3 . A_3$ is therefore the centroid of the points $a_1 . A_1$, $(a_2 + a_3) . B_1$, that is, it is the point G on the straight line A_1B_1, which is determined by the relation

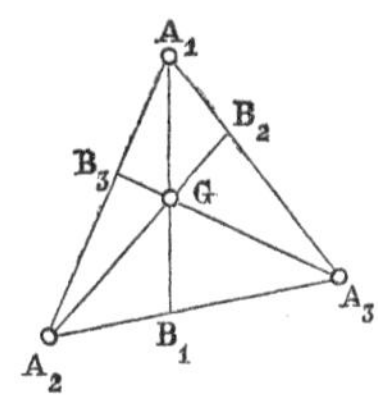

Fig. 104.

$$GB_1 : A_1B_1 = a_1 : a_1 + a_2 + a_3.$$

But the triangles $A_1A_2A_3$, GA_2A_3 are proportional to their altitudes, therefore also to the oblique distances A_1B_1, GB_1 of their vertices from the common base A_2A_3; therefore

$$GA_2A_3 : A_1A_2A_3 = a_1 : a_1 + a_2 + a_3.$$

Similarly we prove that

$$GA_3A_1 : A_2A_3A_1 = a_2 : a_1 + a_2 + a_3,$$
$$GA_1A_2 : A_3A_1A_2 = a_3 : a_1 + a_2 + a_3,$$

and therefore

$$GA_2A_3 : GA_3A_1 : GA_1A_2 = a_1 : a_2 : a_3.$$

That is to say; the centroid G of the three points $a_1 . A_1$, $a_2 . A_2$, $a_3 . A_3$ divides the area $A_1A_2A_3$ into three triangles GA_2A_3,

GA_3A_1, GA_1A_2, which are proportional to the coefficients a_1, a_2, a_3.

Given the points A_1, A_2, A_3, every system of values for the loads a_1, a_2, a_3, determines a point G on the plane $A_1A_2A_3$, and conversely to every point G of the plane there corresponds a fixed system of values equivalent to the above. This is the principle of the calculus of the centroid of Möbius.

126. It follows (from the foregoing articles) that if we wish to find the centroid G of the given points $A_1, A_2, A_3, \ldots$ (Fig. 105) weighted with the coefficients (numbers or segments) a_1, a_2, a_3, ..., we must construct two circuits starting from two different initial points O, O'; the sides of the first being equipollent to $a_1 . OA_1, a_2 . OA_2$, $a_3 . OA_3$, ..., and those of the second to $a_1 . O'A_1$, $a_2 . O'A_2$, $a_3 . O'A_3$, ... The straight lines OR, $O'R'$ which respectively close the two circuits, intersect in the required point G, and we have

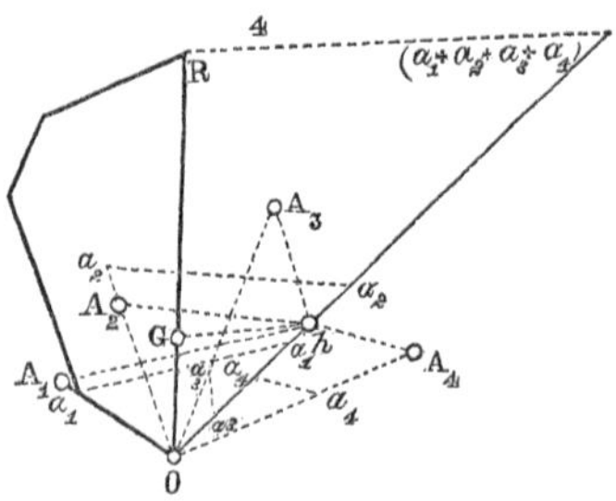

Fig. 105.

$$OR = (a_1 + a_2 + a_3 + \ldots) . OG,$$
$$O'R' = (a_1 + a_2 + a_3 + \ldots) . O'G.*$$

If the coefficients a_1, a_2, &c., &c. are proportional to given segments a_1, a_2, &c., &c. they will also be proportional to loads $\frac{a_1}{h}, \frac{a_2}{h}$, &c., &c. where h is any arbitrary segment; we can therefore make the sides of the first circuit equal to the lengths $\frac{a_1}{h} OA_1, \frac{a_2}{h} OA_2$, &c., &c., and if OR is the closing line then $h . OR = (a_1 + a_2 + a_3 + \ldots) . OG$. Hence it follows, that G is found, without constructing a second circuit, by determining on the closing line the segment

$$OG = \frac{h . OR}{a_1 + a_2 + a_3 + \ldots}.$$

If the coefficients a_1, a_2, &c., ... are proportional to the areas s_1, s_2, &c., &c., which when reduced to a common arbitrary base k are equivalent to the rectangles ka_1, ka_2, ..., &c.; they will

* GRASSMANN, l. c., p. 142.

also be proportional to the segments $a_1, a_2, \ldots$, or to the loads $\frac{a_1}{h}, \frac{a_2}{h}, \ldots$, and if the circuit be constructed with the sides $\frac{a_1}{h} OA_1, \frac{a_2}{h} OA_2$, &c., …, &c., then $OG = \frac{h \cdot OR}{a_1 + a_2 + a_3, \ldots}$.

127. If G is the centroid of the points $a_1 \cdot A_1, a_2 \cdot A_2, a_3 \cdot A_3$, …, and O any point whatever, we have seen that the resultant OR of the segments $a_1 \cdot OA_1, a_2 \cdot OA_2, a_3 \cdot OA_3, \ldots$, is given by the equation

$$OR = (a_1 + a_2 + a_3 + \ldots) \cdot OG,$$

whence

$$OG = \frac{OR}{a_1 + a_2 + a_3 + \ldots}.$$

If $a_1 + a_2 + a_3 + \ldots = 0$, while OR is not zero, then $OG = \infty$, or the centroid G is at an infinite distance. To find in what direction G lies, let B_1 be the centroid of the points $a_2 A_2, a_3 A_3, \ldots, a_n A_n$. Then B_1 is at a finite distance, because $a_2 + a_3 + \ldots, a_n$ is not equal to zero, but is equal to $-a_1$. Let OR be the resultant of $-a_1 OB_1$ and $a_1 OA_1$, this resultant will be equipollent to $a_1 B_1 A_1$, that is it will be independent of the point O. Consequently the resultant of $a_1 OA_1, a_2 OA_2, \ldots, a_n OA_n$, where $a_1 + a_2, \ldots, a_n = 0$, is constant in direction and in magnitude wherever O may be, and is equipollent to the segments $a_1 B_1 A_1 = a_2 B_2 A_2 = \ldots \, a_n B_n A_n$;

where B_r is the centroid of the points $a_1 A_1, a_2 A_2, \ldots, a_{r-1} A_{r-1}, a_{r+1} A_{r+1}, \ldots, a_n A_n$. The point at infinity common to the segments $B_1 A_1, B_2 A_2, \ldots$, is the centroid G of the given points.

Let parallel straight lines be drawn through each of the points $A_1, A_2, \ldots, B_1, B_2, \ldots$ in any arbitrary direction, and let them be cut by a transversal in $A_1', A_2', \ldots, B_1', B_2' \ldots$, the theorem of Art. 121 applied to the points $a_2 A_2, a_3 A_3 \ldots a_n A_n$ and to their centroid B_1 gives

$$a_2 A_2 A_2' + a_3 A_3 A_3', \ldots, a_n A_n A_n' = [a_2 + a_3, \ldots, a_n] \, B_1 B_1'$$
$$= -a_1 B_1 B_1'.$$

Therefore $a_1 A_1 A_1' + a_2 A_2 A_2' + \ldots = a_1 [A_1 A_1' - B_1 B_1']$, consequently $a_1 A_1 A_1' + a_2 A_2 A_2' + \ldots \, a_n A_n A_n' = 0$ if the transversal is parallel to $B_1 A_1$, i.e. is drawn towards the centroid G at infinity.

In the particular case when $OR = 0$, or when B_1 coincides with A_1, B_2 will also coincide with A_2, &c., &c. The centroid G is then quite indeterminate; or, in other words, the system of points $a_1 A_1$, $a_2 A_2$, ... has no centroid. The sum

$$a_1 A_1 A_1' + a_2 A_2 A_2' + \dots, a_n A_n A_n'$$

is then zero, whatever the direction of the parallel lines $A_1 A_1'$, $A_2 A_2'$ ..., and of the transversal $A_1' A_2'$ ... *.

128. Through the points $A_1, A_2, A_3, \dots$, and through their centroid G segments A_1B_1, A_2B_2, A_3B_3, ..., GH are drawn in an arbitrary direction parallel to one another, and proportional to the co-efficients $a_1, a_2, a_3, \dots, m = a_1 + a_2 + \dots$, taking account of signs; that is, having chosen the positive direction of the segments, let the segments proportional to the positive co-efficients be drawn in that direction, and those proportional to the negative coefficients in the opposite direction. Let O be an arbitrary point, and through it draw a straight line parallel to the segments AB, and upon this line project the points $A_1, A_2, A_3, \dots, G$ by parallel rays into $A_1', A_2', A_3', \dots. G'$; then by the theorem of Art. 119 we have

$$a_1 . A_1A_1' + a_2 . A_2A_2' + a_3 . A_3A_3' + \dots = m . GG'.$$

But the numbers $a_1, a_2, a_3, \dots, m$ are proportional to the bases of the triangles OA_1B_1, OA_2B_2, OA_3B_3, ..., OGH, and the segments A_1A_1', A_2A_2', A_3A_3', ..., GG' to the heights of the same triangles, hence the following theorem;

The sum of the triangles which join the segments A_1B_1, A_2B_2, A_3B_3, ..., to O, is equal to the triangle which joins the straight line GH to the same pole O. Whence it follows that GH is the resultant of the segments A_1B_1, A_2B_2, A_3B_3, ..., (Art. 47).

129. This furnishes another construction for the centroid G. After drawing through $A_1, A_2, A_3, \dots$ (Fig. 106) the segments

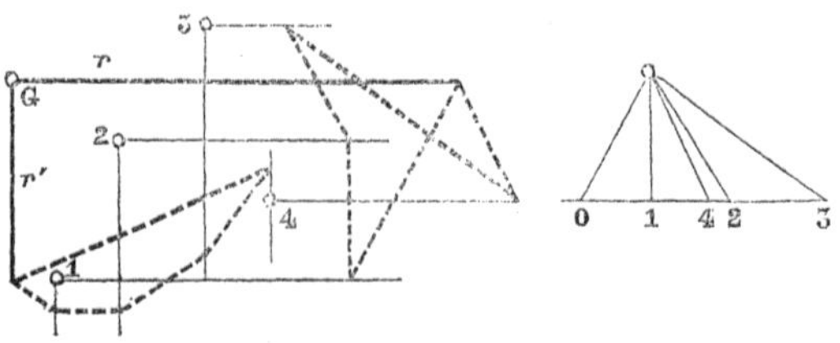

Fig. 106.

$a_1, a_2, a_3, \dots$, in an arbitrarily chosen direction, we combine them in the manner of Article 53.

* MÖBIUS, *Bary. Calcul.*, § 9, 10. BALTZER, *Stereom.*, § 11.

We shall thus obtain a straight line r, in which the resultant segment lies, and which must therefore pass through G. We now repeat this combination, only changing the common direction of the segments $a_1, a_2, a_3, \ldots$, and obtain another straight line r'; the lines r and r' intersect in the required centroid.

130. A figure (linear, superficial, or solid) is called *homogeneous* if all its points are weighted with equal coefficients. Geometrical figures are understood to be homogeneous, unless the contrary is stated.

If the points in a figure are collinear two and two with a fixed point, and situated at equal opposite distances from it, the fixed point is evidently the centroid of the figure. For instance, the centroid of a rectilinear segment is its middle point; the centroid of a parallelogram is the point of intersection of its diagonals; the centroid of a circle, of a circumference, and of a regular polygon, is the geometrical centre of the figure (Figs. 107 and 108).

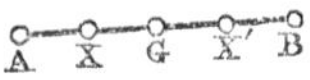

Fig. 107.

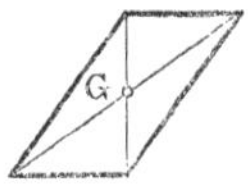

Fig. 108.

If the figure has an axis of symmetry, that is, if its points are two and two on chords bisected normally by an axis, this axis will also contain the centroid.

131. Let the figure be the triangle ABC (Fig. 109). If D is the middle point of BC, the straight line AD divides the area ABC into two equal parts. To every point X in one half there corresponds a point X' in the other half, such that the segment XX' is parallel to BC, and bisected by AD. The centroid of every couple XX' is therefore on AD, hence the centroid G of the area ABC lies on AD. Therefore G is the point of concourse of the three median lines AD, BE, CF. It divides each of the three median lines into two segments which are in the proportion of $2:1$. For, since the triangle ABD is cut by the transversal FGC, we have

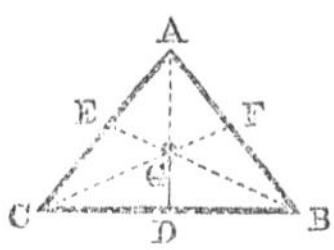

Fig. 109.

$$\frac{AF}{FB}\cdot\frac{BC}{CD}\cdot\frac{DG}{GA} = -1.$$

But $AF = FB$, $BC = 2\,DC$, therefore

$$\frac{DG}{GA} = \tfrac{1}{2},$$

or $GD = \frac{1}{3}AD$, and similarly $GE = \frac{1}{3}BE$, $GF = \frac{1}{3}CF$. The point G is also the centroid of the three points A, B, C.

132. If a (linear or areal) figure is made up of a system of rectilinear segments, or triangular areas, then its centroid is that of the points $a_1.A_1$, $a_2.A_2$, $a_3.A_3$, ..., where A_1, A_2, A_3, ... are the centroids of the segments or triangles, of which the figure is made up, and the (numerical or segmental) coefficients a_1, a_2, a_3, ... are proportional to the segments or triangles themselves.

133. Let the figure be a circuit with rectilinear sides. Let A_1, A_2, A_3, ... be the middle points of the sides, and a_1, a_2, a_3, ... segments proportional to the sides. Then, if we find by one of the methods already described (Articles 126, 129) the centroid G of the points A_1, A_2, A_3, ..., weighted with the segments a_1, a_2, a_3, ...; G is the centroid of the given circuit.

134. If the circuit is part of the perimeter of a regular polygon (Fig. 110), its centroid can be found in a much simpler way. Draw a diameter of the inscribed circle, and let the sides of the circuit be projected orthogonally upon it. Let σ be a side, λ its projection, r the radius of the circle which is drawn through the middle point of σ, and p the perpendicular let fall from the latter point on to the diameter; then the right-angled triangle of which σ is the hypothenuse and λ one

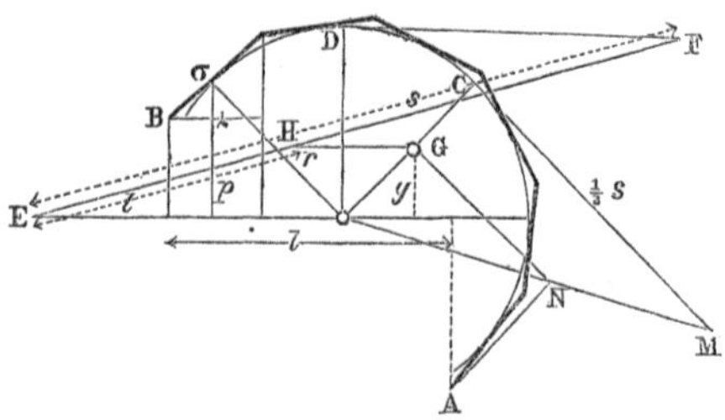

Fig. 110.

of the other sides, is similar to the triangle, whose hypothenuse is r and one of its other sides p.

Therefore we have

$$\frac{\lambda}{p} = \frac{\sigma}{r} \quad \text{or} \quad \lambda r = p\sigma.$$

Write down equations corresponding to these for all the sides of the circuit, and by addition we get

$$rl = p_1\sigma_1 + p_2\sigma_2 + \dots,$$

where l is the projection of the whole circuit.

Let G be the centroid, y the perpendicular let fall from G upon the diameter. Since G is the centroid of the middle points of the sides, supposed to be loaded with the coefficients σ_1, σ_2, &c. respectively, we have (120, 133)

$$p_1\sigma_1 + p_2\sigma_2 + \dots = ys,$$

where s means the length of the whole circuit. Therefore

$$rl = ys, \quad \text{i.e.} \quad y = \frac{rl}{s}.$$

This equation gives the distance of the point G from the diameter; the point G must also lie on that radius (OC) of the circle, which bisects the circuit, since this radius is an axis of symmetry of the circuit. Draw a straight line $EF = s$, of which one extremity E lies on the diameter, and the other extremity F upon the tangent of the circle, which is parallel to this same diameter; and then take upon EF a segment $EH = l$, and through H draw a parallel to DF, cutting the axis of symmetry OC in G; then, since the straight line EF is cut by the parallels EO, HG, DF, we obtain:

$$\frac{EF}{EH} = \frac{\text{distance of } DF \text{ from } EO}{\text{distance of } HG \text{ from } EO},$$

$$\text{or } \frac{s}{l} = \frac{r}{\text{distance of } EO \text{ from } G} = \frac{r}{y},$$

and therefore G is the required centroid. We notice in the formula obtained above, that l is the projection of the (unclosed) circuit upon any diameter chosen at pleasure, and y is the perpendicular distance of the point G from the same diameter.

Another construction. Upon the tangent CM, drawn at right angles to the axis of symmetry OC, set off a segment $CM = \frac{1}{2}s$, join OM, and draw from that extremity (A) of the given circuit, which lies on the same side of the axis of symmetry as M, a parallel to OC, to cut OM in N; through N draw a parallel to CM, till it cuts OC in G. In the similar triangles OCM, OGN the bases are respectively $\frac{s}{2}$, $\frac{l}{2}$, where by l we understand the projection of the circuit upon the diameter perpendicular to OC. The altitude of the first triangle is r, and therefore that of the second is equal to the distance of G from the centre O*.

135. This construction is applicable even when the regular polygon, of whose perimeter the given circuit is a part, has an infinite number of sides, that is, when it becomes a circle. Hence let the given line be an arc AB of a circle whose centre is O (Fig. 111); let s be the length of the arc, the half of which CM is set off along the tangent at its middle point. Project the extremity A into N upon OM by means of a parallel to the axis of symmetry OC, and through N draw a parallel to MC cutting OC in G, then G is the centroid of the arc AB.

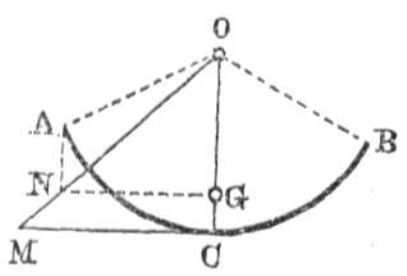

Fig. 111.

For we have

$$CM : CO = GN : GO,$$

$$CM = \tfrac{1}{2}s,\ GN = \tfrac{1}{2}l,\ CO = r,$$

therefore $$GO = y.$$

136. If the given circuit is the perimeter of a triangle ABC (Fig. 112), its centroid G is the centre of the circle inscribed in the triangle DEF, whose vertices are the middle points of the sides of the given triangle. For, D, E, F are the centroids of the rectilinear segments BC, CA, AB; and therefore G is the centroid of the points $\alpha . D, \beta . E, \gamma . F$, where

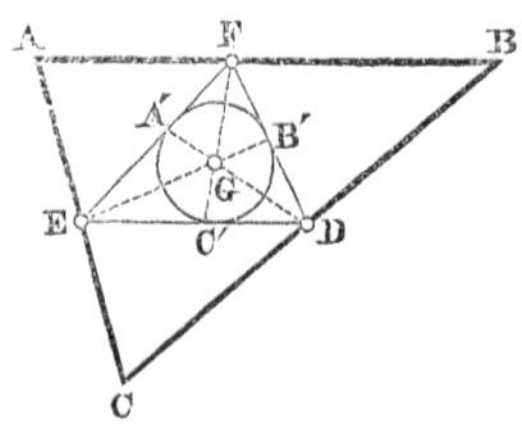

Fig. 112.

$$\alpha : \beta : \gamma = BC : CA : AB.$$

The centroid A' of the points $\beta . E, \gamma . F$ divides the segment EF into two segments EA', $A'F$, such that

$$EA' : A'F = \gamma : \beta = AB : CA = \tfrac{1}{2} AB : \tfrac{1}{2} CA = ED : DF.$$

Therefore DA' is the bisector of the angle EDF, and consequently G, which is the centroid of the points $\alpha . D, (\beta + \gamma) . A'$, lies on the (internal) bisector of the angle D of the triangle

* Culmann, l. c., No. 94.

DEF. Similarly G must also lie upon the bisectors EB', FC' of the other two angles, and therefore G is the centre of the circle inscribed in the triangle *DEF*. Q. E. D.

137. Let the given figure be the quadrilateral *ABCD* (Figs. 113, 114, 115), which may be regarded as the algebraical sum of the two triangles *ABD*, *CDB* into which it is divided by the diagonal *BD*. Let E be the middle point of *BD*. The centroids G_1, G_2 of the two triangles are respectively so situated on the

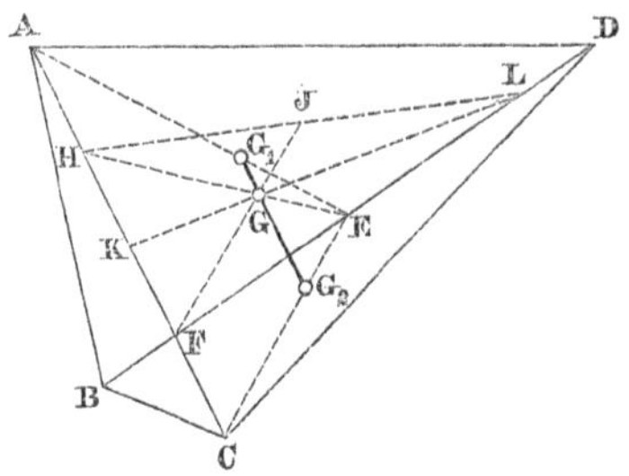

Fig. 113

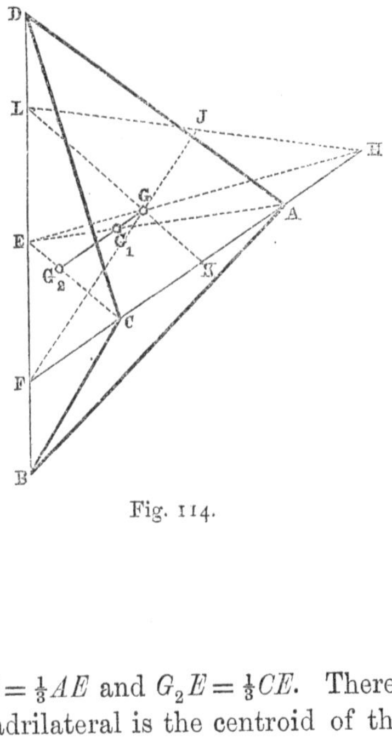

Fig. 114.

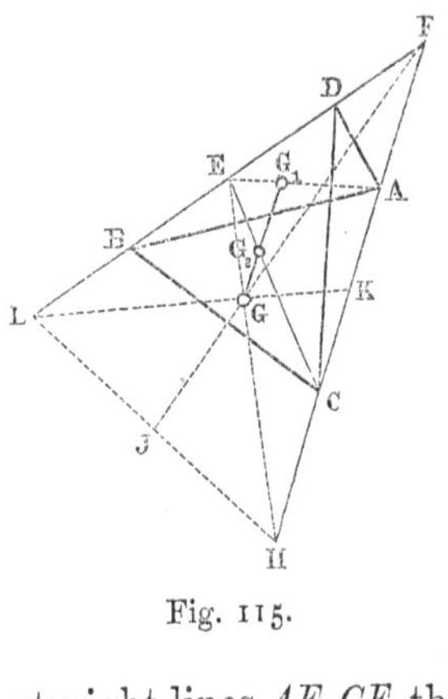

Fig. 115.

straight lines *AE*, *CE*, that $G_1E = \frac{1}{3}AE$ and $G_2E = \frac{1}{3}CE$. Therefore the centroid G of the quadrilateral is the centroid of the two points $a_1 . G_1$, $a_2 . G_2$, where $a_1 : a_2 = ABD : CBD = AF : FC$, where F is the point of intersection of the two diagonals *BD*, *AC*. Since $G_1 G_2$ divides two sides of the triangle *AEC* into proportional parts, it is parallel to the third side *AC*; whence it follows, that the straight line *EG* divides $G_1 G_2$, and *AC* in the same ratio, namely $GG_1 : GG_2 = a_2 : a_1 = FC : AF$. In order to divide *AC* in the ratio $FC : AF$, it is sufficient to interchange the segments AF, FC, that is, to make $AH = FC$, and $HC = AF$. The line joining E to H divides $G_1 G_2$ in the required point G.

The parallels $G_1 G_2$ and AC divide EA, EC, EH in the same ratio; and therefore $GE = \frac{1}{3} HE$, since $G_1 E = \frac{1}{3} AE$, and $G_2 E = \frac{1}{3} CE$.

If instead of BD we employ the diagonal AC, whose middle point is K, and if we interchange the segments BF, FD of BD (i. e. if we take $BL = FD$, and $LD = BF$); then the point G is so situated on LK, that $GK = \frac{1}{3} LK$.

But E, the middle point of BD, is also the middle point of FL, and similarly K is the middle point of FH; hence G is the centroid of the triangle FLH, that is to say:

The centroid of a quadrilateral coincides with that of the triangle, whose vertices are the point of intersection of the diagonals, and the two points obtained by interchanging the segments on each of the two diagonals.

Hence it follows that the straight line FG passes through the middle point J of HL*.

138. If AD, BC are parallel (Figs. 116, 117), and if we draw through the centroids of the triangles BCD, ABD parallels to

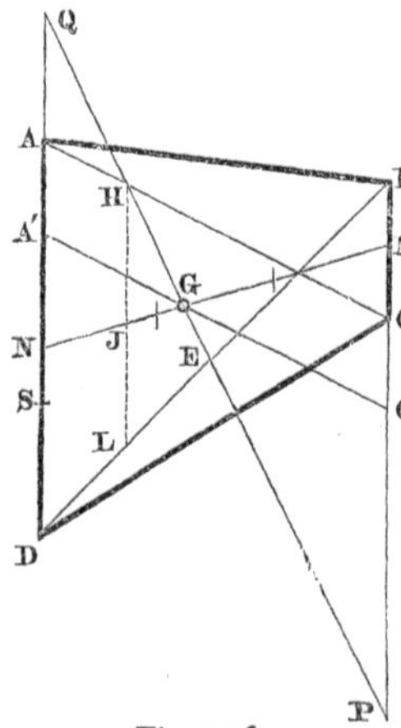

Fig. 116.

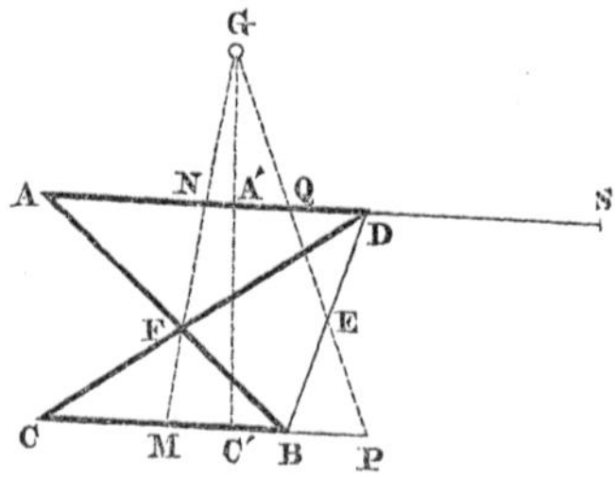

Fig. 117.

AD, these parallels divide the straight line MN which joins the middle points of AD, BC into three equal parts. Since the straight line MN contains the middle points of all

* CULMANN, l. c., No. 95. Cfr. *Quarterly Journal of Mathematics*, vol. 6 (London 1864), p. 127.

chords parallel to AD, it is a diameter of the figure, and therefore the point G lies in it, and divides its central segment into two parts proportional to the areas of the triangles in question, i.e. proportional to BC, AD. The parts of this central segment (since their sum is $\frac{1}{3}MN$, and their ratio $AD:BC$) are respectively equal to

$$\frac{MN.AD}{3(AD+BC)}, \quad \frac{MN.BC}{3(AD+BC)},$$

and consequently

$$MG = \tfrac{1}{3}MN + \frac{MN \times AD}{3(AD+BC)} = \frac{MN(BC+2AD)}{3(AD+BC)},$$

$$GN = \tfrac{1}{3}MN + \frac{MN \times BC}{3(AD+BC)} = \frac{MN(AD+2BC)}{3(AD+BC)};$$

whence $$MG : GN = BC + 2AD : AD + 2BC.$$

Every straight line therefore which passes through G, and is contained between the parallels AD, BC, will be divided by G into two parts proportional to $BC+2AD$ and $AD+2BC$ respectively. If now on BC we take $CP = AD$, and if on AD we take $AQ = CB$, it follows that the straight line PQ will be divided by MN into two segments proportional to MP, QN; but

$$MP = \tfrac{1}{2}BC + AD, \quad QN = BC + \tfrac{1}{2}AD,$$

or $$MP : QN = BC + 2AD : AD + 2BC.$$

Hence PQ passes through G. Since BP, QD are equal and parallel, PQ and BD bisect one another; therefore PQ passes through E the middle point of BD, that is, PQ coincides with HE.

If moreover we take on AD

$$DS = CB, \quad AA' = \tfrac{1}{3}AS,$$

and if on BC we take $CC' = AA'$; then, because

$$A'N = AN - AA' = \tfrac{1}{2}AD - \tfrac{1}{3}(AD - BC) = \tfrac{1}{6}(AD + 2BC)$$

and $$MC' = MC + CC' = \tfrac{1}{2}BC + \tfrac{1}{3}(AD - BC) = \tfrac{1}{6}(BC + 2AD);$$

therefore $$A'N : MC' = AD + 2BC : BC + 2AD,$$

that is $A'C'$ passes through G.

Hence we obtain two simple constructions for the centroid of a quadrilateral with two parallel sides (i.e. a trapezium), either as the intersection of MN with PQ, or as the intersection of MN with $A'C'$*.

* Culmann, *ibid.* Walker, *On an easy construction of the centre of gravity of a trapezium.* (Quarterly Journal of Mathematics, vol. 9, London, 1868, p. 339.)

139. The construction given above for the centroid of a quadrilateral fails in the case where the diagonals AC, BD are parallel (Fig. 118). But in this case the triangles ABD, BCD are equivalent, and of opposite sign, so that $a_1 + a_2 = 0$. It follows that the area of the figure is zero, and the centroid lies at infinity in the direction common to AC and BD.

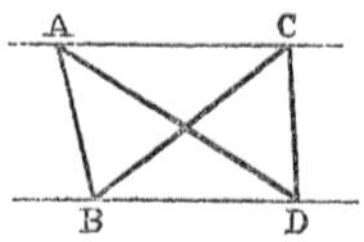

Fig. 118.

140. Now let it be required to find the centroid of any rectilinear figure whatever. We may consider the area of the figure to be the algebraic sum of the triangles, formed by joining the sides of the circuit to an arbitrary point O. Having found the centroids $A_1, A_2, A_3, \ldots$ of these triangles, and reduced their areas to a common base so that they are proportional to the segments $a_1, a_2, a_3, \ldots$, the centroid in question is the centroid of the points $a_1 . A_1, a_2 . A_2, a_3 . A_3, \ldots$ which may be constructed by one or other of the methods already explained.

If the pole O is taken quite arbitrarily, then the number of triangles is equal to the number of sides of the circuit; but if we take O upon one of the sides, or at the point of intersection of two of them, then the number of triangles is reduced by one or two units respectively.

Instead of regarding the proposed figure as the sum of triangles, we may also consider it as the aggregate of the quadrilaterals and triangles, into which it can be decomposed by means of straight lines conveniently drawn.

141. *Example.* Let the given figure be the self-cutting hexagon $ABCDEF$ (Fig. 119), which is the sum of the triangles OBC, OCD, ODE, OFA, O being the point of intersection of the sides AB, EF. Of these four triangles, the first and last are positive, the other two negative. Let their centroids G_1, G_2, G_3, G_4 be found, and let the areas of the triangles, reduced to a common base, be proportional to the segments a_1, a_2, a_3, a_4. These segments a have the same signs as the triangles, the first and last of them are positive, the second and third negative. If now we wish to employ the method of Art. 126, we must first reduce the four products $a_r . OG_r$ to a common base h. In the figure, an arbitrary straight line x is drawn through O, its positive direction is fixed, and upon it the segments h, a_1, a_2, a_3, a_4 are set off from their common initial point O (h, a_1, a_4 in one sense; a_2, a_3 in the opposite sense*). Then the final point of h is joined to G_r, and through the

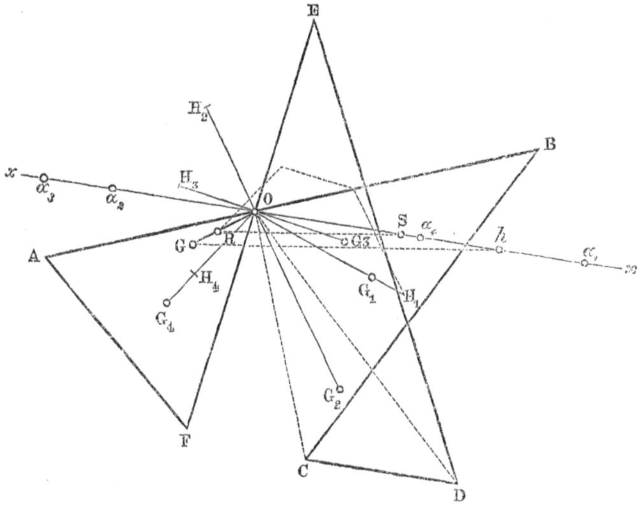

Fig. 119.

final point of ${}_r$ a parallel is drawn to this joining line cutting OG_r in H_r. Thus we obtain $OG_r : h = OH_r : a_r$, and therefore $a_r \cdot OG_r = h \cdot OH_r$. Now construct a circuit starting from O with its sides equipollent to OH_1, OH_2, OH_3, OH_4; the closing line is OR. Finally to construct the point G, given by the relation

$$OG = \frac{h \cdot OR}{a_1 + a_2 + a_3 + a_4},$$

we set off along Ox from its initial point O the segment

$$OS = a_1 + a_2 + a_3 + a_4,$$

join its final point to R, and draw through the final point of h a parallel to this joining line cutting OR in G.

142. Again, let the figure be the cross-section of a so-called *Angle-iron* (Fig. 120). Divide it into six parts, four trapeziums, one triangle, and one parallelogram, denoted in the figure by the numbers 1, 2, 3, 4, 5, 6. Construct the centroids of these six parts, and reduce the areas to a common base, determining the proportional segments 1, 2, 3, 4, 5, 6; and set off these six segments

* In Fig. 119 the final points of the segments h, a are denoted by these letters themselves. Some of the straight lines mentioned in the text are not drawn in the figure.

in succession along a straight line *zz*. Then through an arbitrarily chosen pole *U* draw rays to the points of *zz*, which

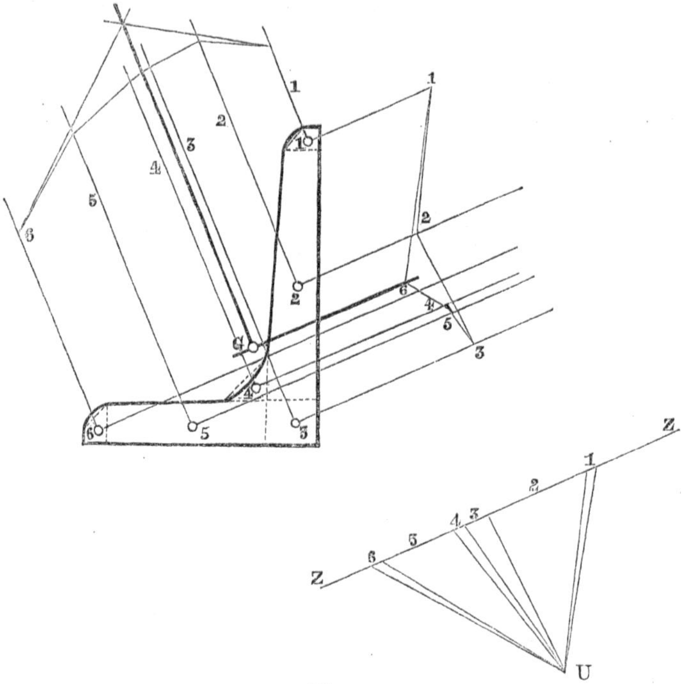

Fig. 120.

bound the segments; next draw through the centroids of the six component figures parallels to *zz*, and construct a polygon, with its vertices lying on these parallels, and its sides respectively parallel to the rays emanating from *U*. The two extreme sides of this polygon will intersect in a point; through which if a parallel to *zz* is drawn, then this straight line must contain the required centroid. In order to obtain a second straight line, possessing the same property, we either repeat the above detailed operations for another direction different to *zz*; or else construct, as shown in the figure, a new polygon, whose vertices lie upon straight lines drawn through the centroids 1, 2, 3, 4, 5, 6 perpendicular to *zz*, and whose sides are respectively perpendicular to the corresponding rays of *U*. It is quite clear that this is just the same thing, as if we drew a new straight line *zz* perpendicular to the first, and then dealt with it just as we formerly dealt with the first *zz*. It should be remembered, that in setting off the segments 1, 2, ... along *zz*, attention must be paid to their signs if the partial areas into which the figure is divided are not all of the same sign *.

143. In the foregoing construction two polygons were used for the purpose of finding two straight lines, passing through the centroid we were in search of. But whenever we know *à priori* one straight line in which the centroid must lie, one polygon is sufficient, for example, when the figure has a diameter. This case is illustrated by the example (Fig. 121), where the figure possesses an axis of symmetry.

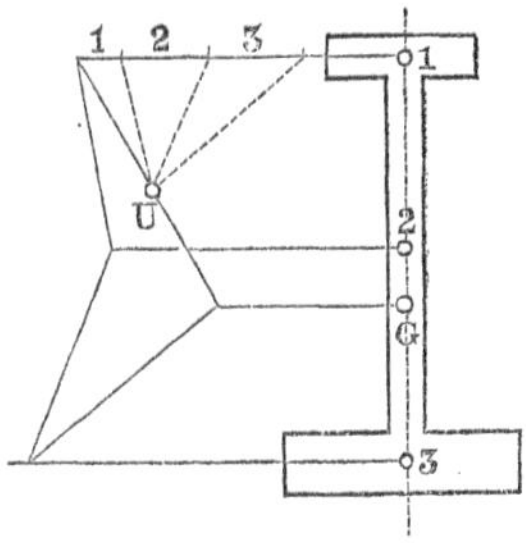

Fig. 121.

The figure represents the cross-section of a *double Tee-iron.*

144. We proceed now to the case of centroids of curvilinear figures, and first we examine that of a circular sector *OAB* (Fig. 122). We consider it to be divided into an indefinitely large number of concentric elementary sectors. The centroid of each of these, regarded as a triangle, lies upon a circle drawn with radius $OA' = \frac{2}{3}OA$. The required centroid is therefore the centroid *G* of the arc *A'B'*. In order to find that point (Art. 135), set off the semi-arc *CA* along the tangent *CM*, join *OM*, and draw *A'N* parallel to *OC* until it intersects *OM* in *N*. Then *G* is the foot of the perpendicular let fall from *N* upon the mean radius *OC* †.

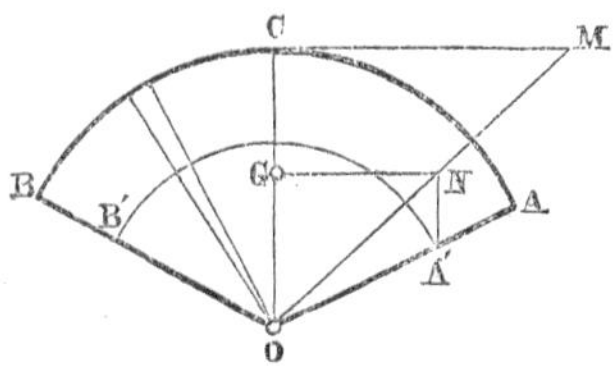

Fig. 122.

145. Next, let the circular segment *ABC* (Fig. 123) be given. This is the difference between the sector *OAB* and the triangle *OAB*, or the sum of the sector *OAB*

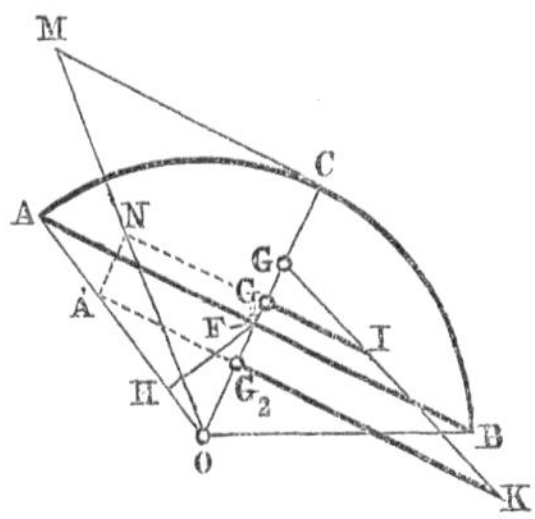

Fig. 123.

* CULMANN, l. c., Nos. 96 & 116.

† CULMANN, l. c., No. 96.

and the triangle OBA. Therefore the centroid G of the segment lies on the straight line (the mean radius OC) joining the centroids G_1, G_2 of the sector and triangle, and divides the segment $G_1 G_2$ into two parts inversely proportional to the areas of these figures. If we take $OA' = \frac{2}{3} OA$, and find the point N as just shown (Art. 144), then G_1, G_2 are the feet of the perpendiculars let fall from N and A' upon the mean radius OC. Let F be the point of intersection of AB and OC, and H the foot of the perpendicular let fall from F upon OA. Then the areas of the sector and triangle are respectively equal to $CM . OA$, and $FH . OA$, that is to say, they are proportional to the lengths CM and FH; therefore, if through G_1 and G_2 two parallel segments $G_1 I$ and $G_2 K$ are drawn in the same sense, equal or proportional to FH, and CM respectively, KI and OC will intersect in G, the required centroid. In fact from the similar triangles $GG_1 I$, $GG_2 K$ we have

$$G_1 G : G_2 G = G_1 I : G_2 K = FH : CM^*.$$

146. If the perimeter of the figure, whose centroid we are finding, consists of rectilinear segments and circular arcs, we decompose the figure by drawing the chords of these arcs or radii to their extremities; then we know how to find the centroid and area of each part, and are able to apply the process of Art. 142.

Example. Let us find the centroid of the figure already dealt with in Art. 107 (Fig. 124). For this purpose we first consider it to be broken up into three parts, the lune, the crown-piece, and the sum of the rectilinear parts; then, regarding the lune as the algebraic sum of two sectors and one quadrilateral, the crown-piece as the algebraic sum of two sectors, and having divided the rectilinear parts by means of the straight line $KCC'K'$, we finally have the given figure equal to the sum of the following parts :

1......Sector $UAEA'$,

2......Quadrilateral $OAUA'$,

3......Sector $AOA'F$,

4......Sector $OB'B$,

5......Sector $OC'C$,

* CULMANN, *ibid.*

6......Trapeziums $BCKH + H'K'C'B'$,
7......Trapeziums $CJIK + K'I'J'C'$.

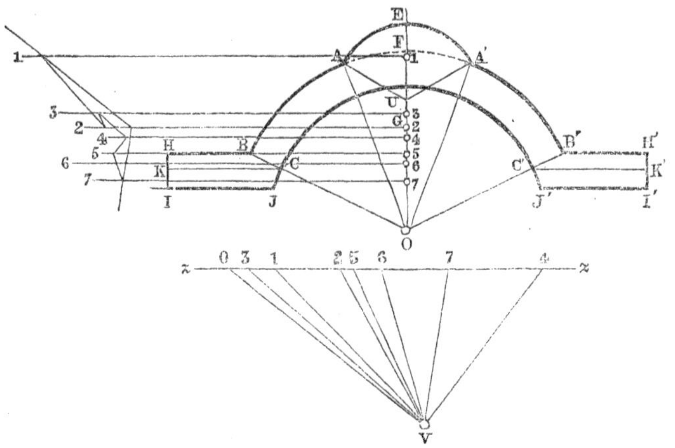

Fig. 124.

We know how to determine the areas of all these, and by reducing them to a common base we are also able to construct their centroids. In order to find the centroid of the sum of $BCKH$ and $H'K'C'B'$, it is sufficient (Art. 138), to find the centroid of the trapezium $BCKH$, and then to draw through it a parallel to KC until it intersects the axis of symmetry EO; the point of intersection is the centroid required. Now to apply the process of Art. 142, we draw, in a direction different to EO, say in that of $KCC'K'$, a straight line zz, on which we set off in succession the segments 1, 2, 3, 4, 5, 6, 7 respectively proportional to the areas of the seven partial figures, noticing that the segments 3 and 5 must be set off in the opposite direction to the others, because they represent negative areas. Through any point whatever V lying outside zz, draw rays to the limiting points of the above segments; then draw lines parallel to zz through the centroids of the partial figures, and construct a polygon whose vertices lie on these parallels, and whose successive sides are parallel respectively to the rays emanating from V. Now draw through the point of intersection of the first and last sides of this polygon a parallel to zz; this line cuts the axis OU in the required centroid of the given figure. This point G falls in our figure very near to the point 2, the centroid of the quadrilateral $OAUA'$. If we produce the sides of the polygon sufficiently, in order to find the point in which the first side cuts the fourth, and also that in which the fourth and sixth intersect, and if through these points we draw parallels to zz till they intersect the axis of symmetry, these latter points of intersection will be the centroids of the lune and the crown-piece.

CHAPTER IX.

RECTIFICATION OF CIRCULAR ARCS.

147. IN order to develope a circular arc AB along its tangent (Fig. 125) we may proceed in the following way. On BA produced mark off a part $AC = \frac{1}{2} BA$, and with C as centre and CB as radius, describe an arc cutting the tangent AD in D. Then AD is the length of the given arc, with a negative error, whose ratio to the whole arc is

$$-\frac{\theta^4}{1080} - \frac{\theta^6}{54432} \ldots,$$

θ being the ratio of the arc to the radius *.

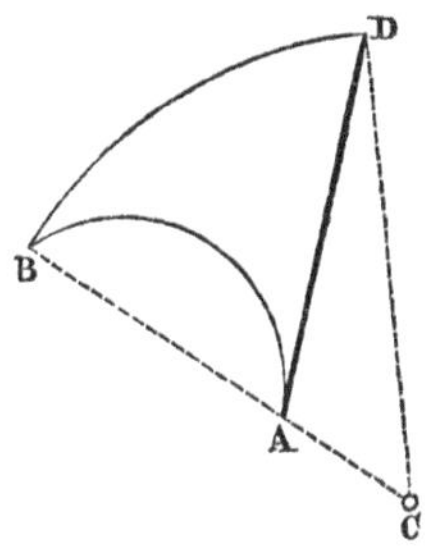

Fig. 125.

Otherwise (Fig. 126): let D be the middle point of the arc AB, and E the middle point of the arc AD; let the radius OE intersect the tangent at A in C, and join CB; then $AC + CB$ is the length of the given arc with a positive error, whose ratio to the whole arc is

$$+\frac{\theta^4}{4320} + \frac{\theta^6}{3484648} \ldots.$$

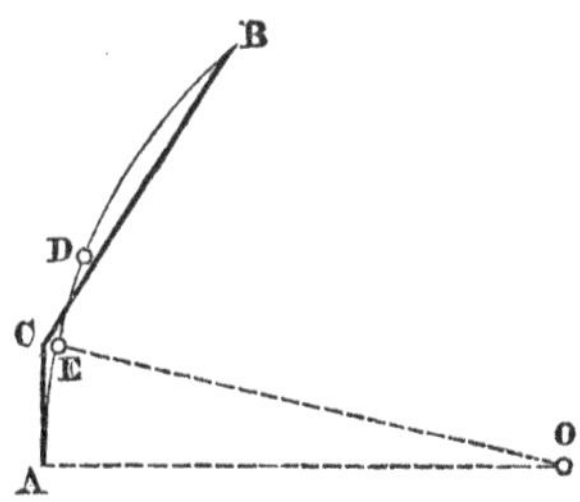

Fig. 126.

Since $4320 = 4 \times 1080$, if we add to $\frac{4}{5}$ of the length found by the second construction $\frac{1}{5}$ of that found by the first, the sum obtained will be very approximately equal to the length of the arc, with a positive error, whose ratio to the whole length of the arc is

$$+\frac{17\theta^6}{870912} \ldots †.$$

* RANKINE, *On the approximate drawing of circular arcs of given length* (Philosophical Magazine, October, 1867), p. 286.

† RANKINE, *On the approximate rectification of circular arcs* (Philosophical Magazine, November, 1867), p. 381.

For the proof of these rules we refer the reader to the original memoirs of Professor Rankine, cited in the foot-notes.

148. In regard to this question, it will be convenient to mention at this point some methods suggested by Professor A. Sayno, of Milan.

The method given by Culmann for developing a circular arc AB along the tangent at one of its points is much too long. The length of a circular arc may be found graphically in a much simpler fashion, by having recourse to auxiliary curves, which drawn once for all can be employed in every example.

Consider a convolution $OMRS$ of the *Spiral of Archimedes*, which when referred to its polar axis OX and its pole O, has

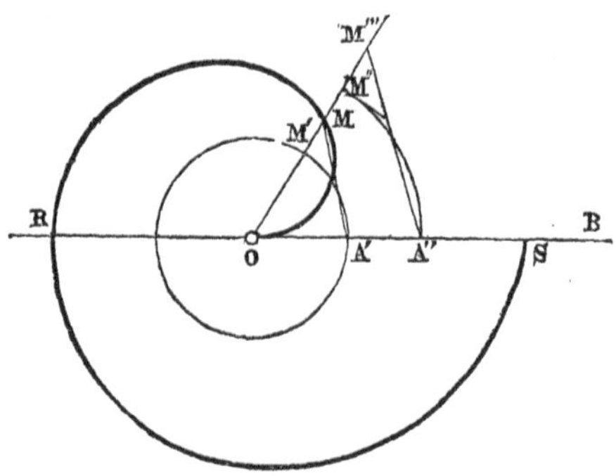

Fig. 127.

the equation $\rho = a\ \omega$ *, and the circle drawn with centre O and radius $OA' = a$. Let OM be any radius vector of the spiral, which cuts the circle in M'; then the arc $A'M' = OM$. If now we wish to find the length of an arc $A''M''$ of any radius whatever OA'', it is sufficient to place the spiral (supposed moved from its previous position) so that its polar axis coincides with the radius OA'' of the given arc, to mark on OA'' the point A', and on the other radius OM'' the point M in which it cuts the curve. Now take the spiral away and draw through A'' a parallel to $A'M$, cutting OM'' in M''', then OM''' is the required length of the arc. We can construct this spiral upon a thin plate of brass, horn, or ivory; it is sufficient

* ρ is the radius vector OM, and ω the corresponding vectorial angle $A'OM$.

to mark upon it the pole and the point A'. This would be a new instrument, which might be added as a 'Graphometer' to the case of drawing implements of an Engineer.

The *Spiral of Archimedes* $\rho = a\omega$ (Fig. 128) enables us also to develope the arc along the tangent. Having drawn the

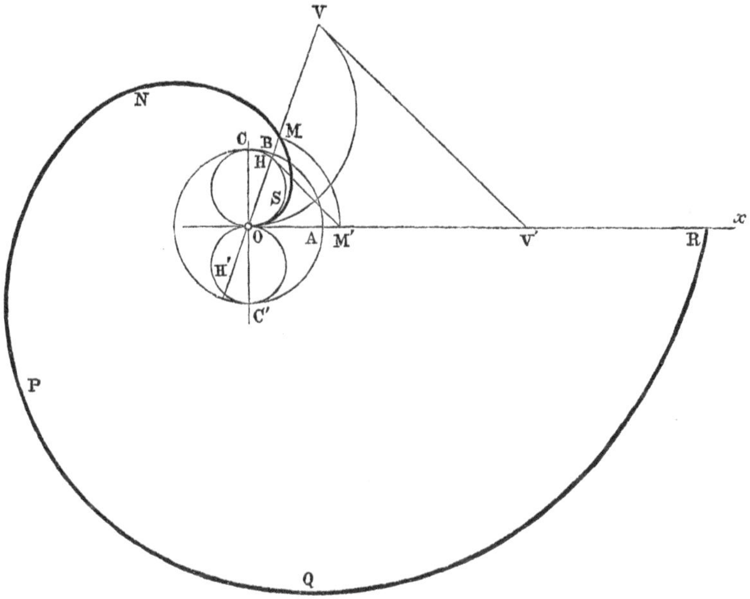

Fig. 128.

circle whose radius $OA = a$, and the circle whose diameter $OC = OA$, if B, H are the points in which these circles are cut by any radius vector OM, then $OM =$ the arc $AB =$ arc OH. Therefore, if we wish to set off the arc OV along the tangent OX, we need only place the spiral in such a manner that the pole and the polar axis coincide respectively with the point of contact O and the tangent OX of the given arc, and then mark the points H, M in which the chord OV cuts the circle on OC as diameter, and the spiral. We then take away the spiral, and mark off on OX the segment $OM' = OM$; draw through V a parallel VV' to HM', and OV' is the required length of the arc.

In order to increase the stiffness of the plate which forms the instrument, it is best to use the circle of radius

$OC' = OC$, and then, supposing the chord VO to be produced, we obtain $OH' = HO$.

Another curve, which serves the same purpose, is the *hyperbolic spiral*, whose equation in polar coordinates is $a = \rho\,\omega$. Draw (Fig. 129) a convolution of this curve $NMDCBA$, and

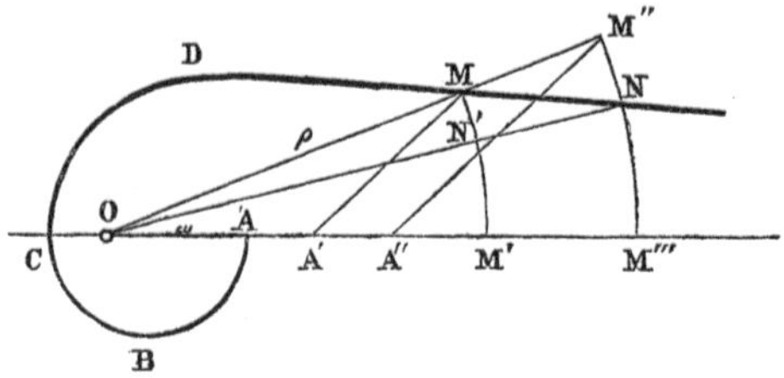

Fig. 129.

mark off a point A' on the polar axis, such that $OA' = a$. Then the length of the circular arc MM', of radius OM, is OA'; hence the length of any circular arc whatever $M''\,M'''$, drawn with its centre at O, is OA'', where A'' is got by drawing $M''\,A''$ parallel to MA'. This curve however is of no use in determining the lengths of small arcs, so that for practical purposes the first curve is to be preferred.

The hyperbolic spiral enables us also to divide angles in a very elegant manner. Thus, to find the arc $M'N' = \frac{1}{n}\,M'M$ (Fig. 129), we need only produce the radius vector OM, take $OM'' = n \cdot OM$, and draw an arc of radius OM'' to cut the spiral in N; the radius ON meets the arc $M'M$ in the required point N'.

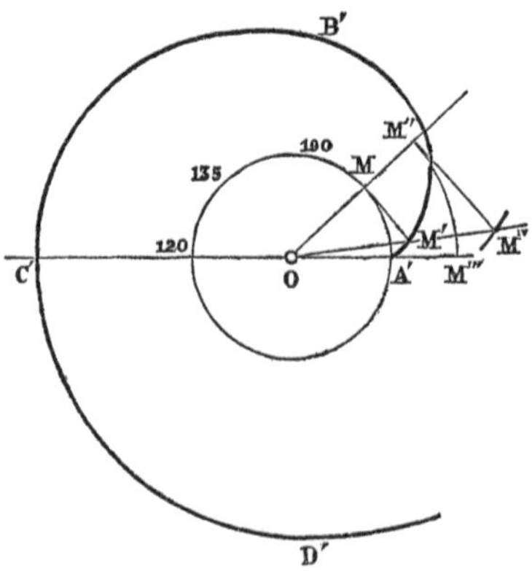

Fig. 130.

In order to set off the arc along the tangent we can also employ another auxiliary curve, namely the involute of the circle. Take (Fig. 130) a circle of radius OA', and let $A'M'B'C'D'$ be its involute. From the figure we have at once the arc $MA' = MM'$, where MM' is the tangent of the circle at M. If now it is required to set off the arc $M''M'''$ (whose centre is O), along its tangent from M'', we need only draw OM', which if sufficiently produced cuts the tangent in question in M^{IV}, and $M''M^{IV}$ is the required length of the arc.

149. By far the simplest method of rectifying the semi-circumference is that of a Polish Jesuit, Kochansky, which was published in the Acta Eruditorum Lipsiæ, year 1685, page 397, according to Dr. Böttcher*. Let O be the centre and

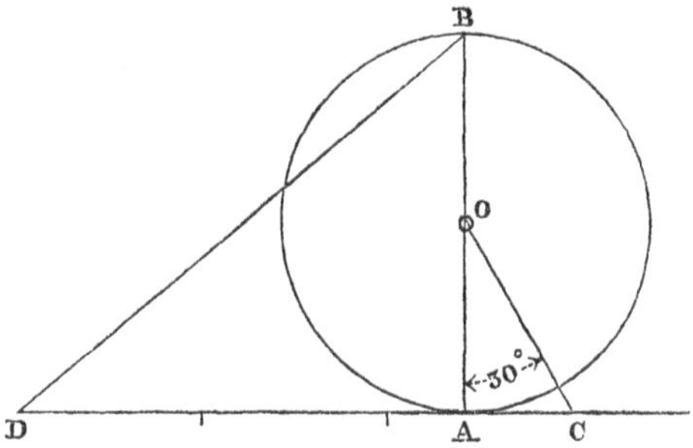

Fig. 131.

AB a diameter of the circle of radius = 1, the angle $COA = 30°$. Then if we take CD = three times the radius, we have

$$\overline{BD}^2 = \overline{BA}^2 + (CD - CA)^2 = 4 + (3 - \tan 30°)^2,$$

i. e. $$BD = 3{\cdot}14153,$$

a value of the semi-circumference true to four places of decimals.

By means of this method, the rectification of an arc greater than 90° can be reduced to the rectification of its supplementary arc.

* [In the XVI vol. (Leipsic, 1883) of Hoffmann's Zeitschrift für math. und naturl. Unterricht.]

RECIPROCAL FIGURES IN GRAPHICAL STATICS.

AUTHOR'S PREFACE

TO THE ENGLISH EDITION.

At a time when it was the general opinion that problems in engineering could be solved by mathematical analysis only, Culmann's genius suddenly created Graphical Statics, and revealed how many applications graphical methods and the theories of modern (projective) geometry possessed.

No section of Graphical Statics is more brilliant or shows more effectually the services that geometry is able to render to mechanics, than the one dealing with reciprocal figures and framed structures with constant load.

It is to this circumstance that I owe the favourable reception my little work (*Le figure reciproche nella statica grafica*, Milano, 1872) met with everywhere; and not the least from Culmann himself. It has already had the honour of being translated into German and French. Having been requested to allow an English version of it, to be published by the Delegates of the Clarendon Press, I consented with pleasure to Professor Beare undertaking the translation.

I have profited by this occasion to introduce some improvements, which I hope will commend themselves to students of the subject.

L. CREMONA.

Rome, *October*, 1888.

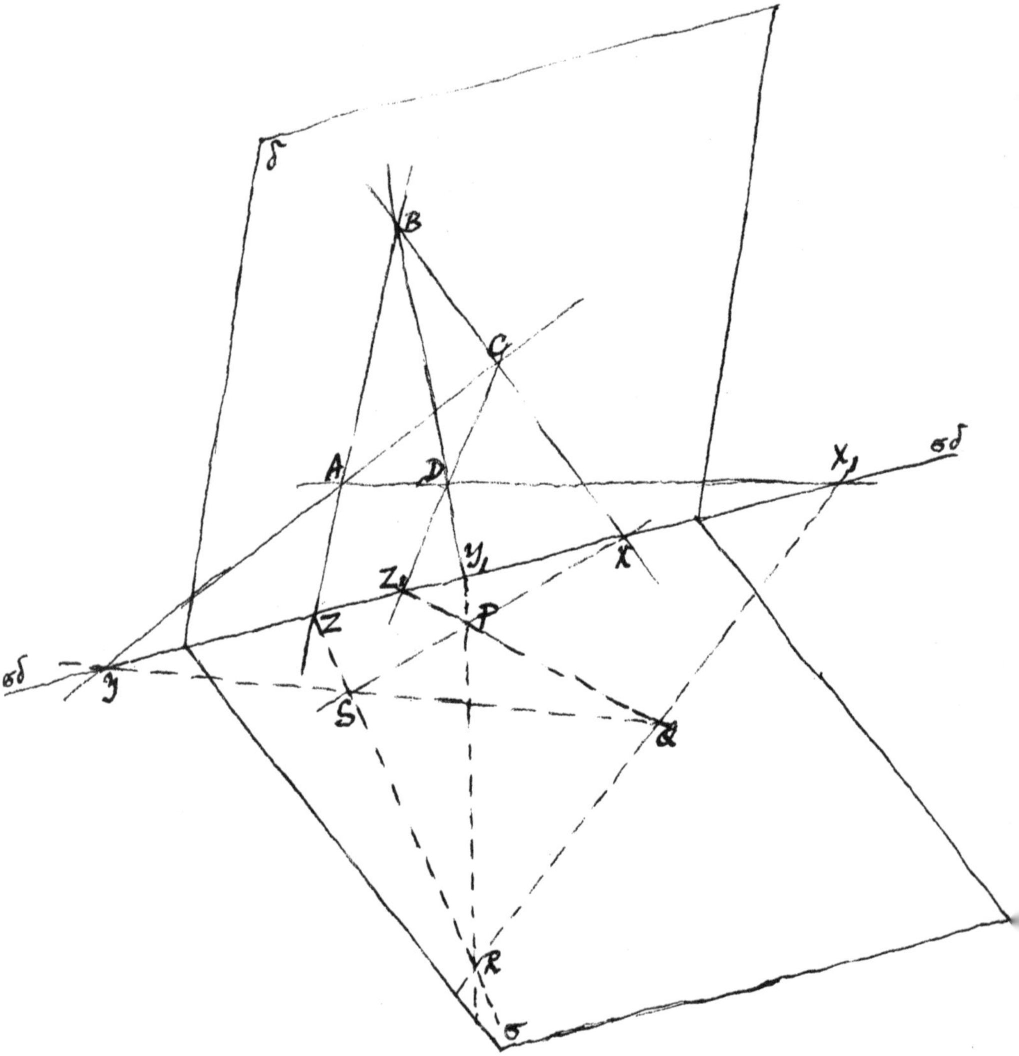

δ
B
C
A
D
X₁
σδ
y₁
X
Z₁
Z
P
σδ
Y
S
Q
R
σ

CHAPTER I.

POLE AND POLAR PLANE.

1. THAT dual and reciprocal correspondence between figures in space, discovered by Möbius*, in which, to any plane whatsoever, corresponds a pole situated in the same plane, and all planes passing through any one point have their poles on the polar plane of that point is called a *Null-system* by German mathematicians.

Such a correspondence is obtained in the following manner. Let there be a plane δ, and four points in it A, B, C, D, no three of which are in one straight line; and let there be three other planes α, β, γ passing through AD, BD, CD, respectively. These will be the fixed elements in the construction.

Draw any plane whatever σ cutting the straight lines $\beta\gamma$, $\gamma\alpha$, $\alpha\beta$ in P, Q, R respectively, then the planes PBC, QCA, RAB will all intersect in the same point S of the plane σ.

Demonstration. Let X, Y, Z, X_1, Y_1, Z_1 be the points in which the straight line $\sigma\delta$ intersects the sides BC, CA, AB, AD, BD, CD of the complete quadrilateral $ABCD$; these points form three pairs of conjugate points of an involution†, by Desargues's Theorem. Since the planes δ, σ, α, meet in X_1 the straight line QR common to the planes σ, α passes through that point; similarly RP passes through Y_1, and PQ through Z_1. Of the six points in involution, taken now in the plane σ, three, X_1, Y_1, Z_1, belong to the sides QR, RP, PQ of a triangle PQR; therefore‡ the straight lines XP, YQ, ZR meet in one point S, which with PQR forms a complete quadrilateral.

* MÖBIUS, *Ueber eine besondere Art dualer Verhältnisse zwischen Figuren in Raume*, in vol. x. of Crelle's Journal, Berlin, 1833, or in vol. i. p. 489, *Gesammelte Werke*, Leipzig, 1883.

In reality this system of reciprocal figures in space had been already discovered by GIORGINI (1827), (Memorie della Società Italiane delle Scienze Modena, vol. xx).

† CREMONA, *Projective Geometry* (Oxford, 1885), Art. 131.

‡ CREMONA, *Projective Geometry* (Oxford, 1885), Art. 135.

Therefore the planes $BCPX$, $CAQY$, $ABRZ$ meet in a point S of the plane PQR.

This theorem may be expressed as follows.

If the faces of a tetrahedron $ABCS$ pass respectively through the vertices of another tetrahedron $PQRD$, and if three faces of the latter pass through three vertices of the former, then the fourth face of the second tetrahedron will pass through the fourth vertex of the first (Theorem of Möbius*).

2. Starting from the fixed elements A, B, C, α, β, γ, let any plane σ whatever be given, and let it be required to determine by means of this theorem the point S lying in it.

The plane σ meets the straight lines $\beta\gamma$, $\gamma\alpha$, $\alpha\beta$ in three points P, Q, R, and the three planes PBC, QCA, RAB intersect in the required point S.

Conversely, given any point S whatever, to determine the corresponding plane σ, which passes through S.

The planes SBC, SCA, SAB intersect $\beta\gamma$, $\gamma\alpha$, $\alpha\beta$ in three points P, Q, R, the plane of these points is the required plane.

The point S is called the *pole* of the plane σ, and the latter is termed the *polar plane* of S.

3. If the plane σ change its position, the points Q, R in it remaining fixed, the planes QCA, RAB will remain fixed, and therefore the point S will move on the straight line (which passes through A) common to these two planes. When the point P falls on D, that is, when σ coincides with QRD (i.e. α), the plane PBC coincides with $ABCD$, and S falls on A. Then A is the pole of the plane α, and similarly B and C are the poles of β, γ.

If the arbitrary plane σ passes through BC, the traces of the planes QCA, RAB on it, will be the straight lines QC, RB which are the traces of the planes γ, β; therefore the pole falls in the straight line $\beta\gamma$, i.e. on P. The points P, Q, R are consequently the poles of the planes PBC, QCA, RAB.

If the arbitrary plane coincides with ABC, the point P falls on D, i.e. D is the pole of the plane ABC.

* Möbius, *Kann von zwei dreiseitigen Pyramiden eine jede in Bezug auf die andere um und ein-geschreiben zugleich heissen?* vol. iii. of Crelle's Journal (Berlin, 1828), or *Gesammelte Werke*, vol. i. p. 439.

4. The pole S of the arbitrary plane σ (or conversely the polar plane σ of the arbitrary point S) has been determined starting from the system, supposed given, of three planes α, β, γ (having no straight line in common) and their poles A, B, C. But in the tetrahedron $ABCS$ the relations between the vertices (or the faces) are perfectly reciprocal, that is, are interchangeable; so that just as S has been deduced from $ABC\alpha\beta\gamma$, so A may be determined from $SBC\sigma\beta\gamma$; and so on. From this it follows that if S_1, S_2, S_3 are the poles of any three arbitrary planes σ_1, σ_2, σ_3 (not passing through the same straight line), deduced in the manner above described from the system $ABC\alpha\beta\gamma$, the pole S of the plane σ, determined from this same system, coincides with that which would be determined by a similar construction starting from $S_1S_2S_3\sigma_1\sigma_2\sigma_3$ as the given system.

5. From the theorem of Möbius it follows that if the plane σ be drawn through the pole P of a plane $\pi = PBC$, the pole S of the plane σ falls in π; therefore :—

If a plane passes through the pole of another plane, conversely the latter contains the pole of the former, that is to say :—

If a point lies in the polar plane of a second point, the latter lies in the polar plane of the former.

From this it follows that the poles of all the planes passing through a point S lie in a single plane σ, the polar plane of S; and the polars of all the points of a plane σ pass through one and the same point S, the pole of σ.

6. Let α, β be two planes, and A, B their poles. Any plane whatever through AB will have its pole in α and in β, that is, in the straight line $\alpha\beta$; conversely, the polar plane of any point whatever of $\alpha\beta$ will pass through A and B, i.e. through the straight line AB. And any plane whatever through the straight line $\alpha\beta$, which contains the poles of the planes through AB, will have its pole on the straight line AB; and conversely, any point whatever of AB, being on the polar plane of the points of $\alpha\beta$, will be the pole of a plane through $\alpha\beta$.

Two straight lines, such as $\alpha\beta$ and AB, each of which is the locus of the poles of planes passing through the other, are called *reciprocal straight lines.*

Hence it follows that if a straight line r passes through a point A, its reciprocal r' lies in a, the polar plane of A; and conversely.

7. A straight line r, which lies in a plane a and passes through A, the pole of a, coincides with its reciprocal, that is to say, it is reciprocal to itself. In fact, if M is any other point whatever of r, since M lies in a, the polar plane of A, then μ, the polar plane of M, passes through A. And since μ must also pass through M, the polar plane of it or of any point whatever of the given straight line, r passes through the straight line r.

From this it follows, that two reciprocal straight lines r and r', which are non-coincident, cannot lie in one plane. If a plane a passes through both r and r', the pole A of the plane will be on both r and r', and r would lie in a plane and contain its pole, therefore r would be reciprocal to itself.

All straight lines reciprocal to themselves and passing through a given point A lie in a, the polar-plane of A. All straight lines reciprocal to themselves and lying in a given plane a pass through A, the pole of a.

A system of straight lines reciprocal to themselves is called a *linear complex*, and the straight lines are called *rays* of the complex.

Each ray of the complex which meets a given straight line r (not itself a ray) meets also its reciprocal straight line r'. In fact, if A is the point common to the ray and to r, the plane a, the polar of A, must pass through the ray and the straight line r'.

Conversely, if a straight line t meets two reciprocals r and r', the straight line t is necessarily a ray. For, the point tr is the pole of a plane which passes through this point, and through r'; therefore the plane also passes through t. Hence t lies in a plane polar to one of its own points, or t is a ray.

From this it follows that all the straight lines (necessarily rays) cutting two reciprocal straight lines r and r', and another line s, also meet the straight line s' reciprocal to s. Two pairs rr', ss' of reciprocal straight lines are therefore situated on the same hyperboloid, the generators of which are all rays of the complex of another system.

8. All planes parallel to the same plane may be considered * as having in common a line r' situated at infinity, therefore their poles all lie on a straight line r, the reciprocal of r'. Changing the bundle of parallel planes, the straight line r remains parallel to itself, because it passes through a fixed point I lying at infinity, that is, through the pole of the plane ι at infinity, in which the straight line r' is always situated.

Such lines r, whose reciprocals lie at infinity, are called *diameters* of the complex.

Planes perpendicular to the common direction of the diameters are parallel to each other, therefore their poles are on a diameter. This diameter a, which is distinguished from the other diameters by being perpendicular to the planes whose poles it contains, is called the *central axis* of the complex.

Straight lines parallel to the central axis are reciprocals to straight lines in the plane at infinity ι; and in particular the central axis is reciprocal to the line at infinity common to all planes perpendicular to the central axis itself. The point I, at infinity on the central axis, is the pole of the plane at infinity.

9. If r and r' are any two reciprocal straight lines whatever, the straight line which passes through their points at infinity will be a ray of the complex, and will therefore pass through the pole I of the plane at infinity; that is, the points at infinity of two reciprocal straight lines and of the central axis are all three in one straight line. Hence it appears that two reciprocal straight lines and the central axis are parallel to the same plane.

Therefore, planes parallel to the central axis and passing through two reciprocal straight lines are parallel to each other.

From this it follows that:

If two reciprocal straight lines are projected parallel to the central axis, on a plane, not containing the direction of the central axis, their projections will be two parallel straight lines.

We shall suppose that the projection is made on a plane perpendicular to the central axis.

* Cremona, *Projective Geometry* (Oxford, 1885), Art. 26.

Moreover, it follows that the straight line meeting two reciprocal straight lines and perpendicular to them cuts the central axis orthogonally.

10. Suppose the central axis horizontal, and let us call that plane of projection which intersects the central axis in its own pole the *orthographic plane.*

Take that point as the origin of a system of rectangular coordinates x, y, z, and let the axis of z coincide with the central axis: then the preceding theorems and laws of reciprocity will be expressed by the following equations.

The point (x_1, y_1, z_1) is the pole of the plane

$$xy_1 - yx_1 + k(z - z_1) = 0$$

where k is some constant.

Conversely the plane

$$ax + by + cz + d = 0$$

corresponds to the pole

$$x = -\frac{kb}{c}, \quad y = \frac{ka}{c}, \quad z = -\frac{d}{c}.$$

The straight line

$$\begin{aligned} ax + by + c &= 0 \\ px + qy + rz &= 0 \end{aligned}$$

is reciprocal to the straight line

$$\begin{aligned} ax + by + c' &= 0 \\ px + qy + r'z &= 0 \end{aligned}$$

where $rc' = r'c = k(aq - bp)$.

11. Hence:

(*a*) *To any number of straight lines r in space, the projections of which coincide in a single straight line, straight lines r′ correspond, whose projections are coincident or parallel, according as the straight lines r (necessarily lying in a plane parallel to the central axis) are parallel or not.*

(*b*) *To any number of straight lines r in space, the projections of which are parallel, straight lines r′ correspond, whose projections are coincident or parallel, according as the straight lines r (necessarily parallel to a plane passing through the central axis) are parallel or not.*

12. If the points $A, B, C, D \ldots$ in space are considered as vertices of a polyhedron, the polar planes $\alpha, \beta, \gamma, \delta, \ldots$ are the faces of a second polyhedron, whose vertices $\alpha\beta\gamma, \ldots$ are the poles of the faces $ABC, \ldots$ of the first. The two polyhedra are called *reciprocal*; to the vertices of each correspond the faces of the other, to the edges the edges. Each polyhedron is simultaneously inscribed and circumscribed to the other (Art. 1). Two corresponding edges are reciprocal straight lines (Art. 6).

Let the two polyhedra be projected on the orthographic plane; the projections will be two figures possessing reciprocal properties. To each side of the first figure there will correspond a parallel side of the other, since two corresponding sides are the projections of two reciprocal edges of the two polyhedra. If one of the polyhedra has a solid angle, at which m edges meet, the other will have a polygonal face of m sides; and therefore, if in one of the orthographic figures there are m sides diverging from a point or *node*, the m corresponding sides of the other orthographic figure will be the sides of a closed polygon.

In a polyhedron, each edge is common to two faces, and joins two vertices; each face has at least three edges, and in each vertex at least three edges meet; hence in both orthographic figures, each side is common to two polygons, and joins two nodes, three sides at least meet in every node, and each polygon has at least three sides.

Suppose that one of the polyhedra, and consequently the other, belongs to the class of Eulerian polyhedra*; then the sum of the numbers of vertices and faces exceeds by two the number of edges, from the well-known theorem of Euler. Hence, if the first orthographic figure possesses p nodes, p' polygons, and s sides, we have

$$p+p'=s+2.$$

The second figure will have p' nodes, p polygons, and s sides.

13. If one polyhedron has a vertex at infinity, the other has a face perpendicular to the orthographic plane, and conversely; consequently, if one of the orthographic figures has a vertex at infinity, the other contains a polygon whose sides all lie in the same straight line, and conversely.

If the point I at infinity on the central axis is a vertex common to n faces of the first polyhedron, then the other

* TODHUNTER, *Spherical Trigonometry*, Chapter xiii, Polyhedra.

polyhedron has in the plane at infinity a polygonal face of n sides. In this case, the first orthographic figure has $p-1$ nodes, $p'-n$ polygons, and $s-n$ sides; and the second (not reckoning the straight line at infinity) possesses $p-1$ polygons, $p'-n$ nodes, and $s-n$ sides: where the numbers p, p', s are still connected by the relation

$$p+p' = s+2.$$

CHAPTER II.

POLYGON OF FORCES AND FUNICULAR POLYGON AS RECIPROCAL FIGURES.

14. THOSE *reciprocal diagrams*, which are obtained as the orthographic projections of two reciprocal polyhedra, present themselves directly in the study of *graphical statics*. The mechanical property of reciprocal diagrams is expressed in the following theorem due to the late Professor Clerk Maxwell*:

'*If forces represented in magnitude by the lines of a figure be made to act between the extremities of the corresponding lines of the reciprocal figure, then the points of the reciprocal figure will all be in equilibrium under the action of these forces.*'

The truth of the theorem is at once apparent, if we observe that the forces applied at any node whatever of the second diagram are parallel and proportional to the sides of the corresponding closed polygon of the first diagram.

The theorem is particularly useful, in the graphical determination of the stresses, which are developed in frame-work structures.

15. The first germs of the theory are met with in the properties of the *polygon of forces*, whose sides represent in magnitude and direction a system of forces in equilibrium applied at any point; and also in the well-known geometrical constructions which enable us to determine the tensions of the sides of a plane funicular polygon †. But the first to apply the theory to frame-work structures was the late Professor Macquorn Rankine, who, in Art. 150 of his excellent work *Manual of Applied Mechanics* (1857), proved the following theorem:

'*If lines radiating from a point be drawn parallel to the lines of resistance of the bars of a polygonal frame, then the sides of any*

* Philosophical Magazine, April 1864, p. 258.

† VARIGNON, *Nouvelle Mécanique ou Statique, dont le projet fut donné en* 1687: Paris, 1725.

K 2

polygon whose angles lie in these radiating lines will represent a system of forces, which, being applied to the joints of the frame, will balance each other; each such force being applied to the joint between the bars whose lines of resistance are parallel to the pair of radiating lines that enclose the side of the polygon of forces, representing the force in question. Also, the lengths of the radiating lines will represent the stresses along the bars to whose lines of resistance they are respectively parallel.'*

Rankine afterwards published an analogous theorem for a system of polyhedral frames †.

16. The geometrical theory of reciprocal diagrams is specially due to the late Professor Clerk Maxwell, who first in 1864 ‡, and again in 1870 §, defined them generally, and obtained them from the projections of two reciprocal polyhedra.

But his polyhedra are *reciprocal in respect to a certain paraboloid of revolution*, in the sense of the *theory of reciprocal polar figures of Poncelet* ‖; so that, projecting orthogonally and parallel to the axis, the corresponding sides of their projections are not parallel, but perpendicular to one another. Hence we must rotate one of the diagrams through $90°$ in its own plane, in order that it may assume that position which it ought to take in statical problems.

On the contrary, by the more general process, explained in this treatise, the orthographic projections of two reciprocal polyhedra give precisely those diagrams which occur in graphical statics.

17. The practical application of the method of reciprocal figures was made the subject of a memoir by the late Professor Fleeming Jenkin, communicated in March 1869 to the Royal Society of Edinburgh ¶. In that memoir, after quoting

* Page 142 of the sixth edition (1872).

† Philosophical Magazine, Feb. 1864, p. 92.

‡ *On reciprocal figures and diagrams of forces* (Philosophical Magazine, April 1864, p. 250).

§ *On reciprocal figures, frames and diagrams of forces* (Transactions of the Royal Society of Edinburgh, vol. xxvi. p. 1). See also a letter of Professor Rankine in the '*Engineer*,' Feb. 1872.

‖ Or, rather, that of Monge. (See Chasles, *Aperçu historique*, p. 378.)

¶ *On the practical application of reciprocal figures to the calculation of strains on framework* (Transactions of the Royal Society of Edinburgh, vol. xxv. p. 441). See also, by the same author: *On braced arches and suspension bridges,* the definition of reciprocal figures, and their statical property, as enunciated by Maxwell in his memoir of 1864, he adds:

'*Few engineers would, however, suspect that the two paragraphs quoted put at their disposal a remarkably simple and accurate method of calculating the stresses in framework; and the author's attention was drawn to the method chiefly by the circumstance that it was independently discovered by a practical draughtsman, Mr. Taylor, working in the Office of the well-known contractor Mr. J. B. Cochrane.*'

He also presents several examples, accompanied by figures, and finishes with this observation:

'*When compared with algebraic methods, the simplicity and rapidity of execution of the graphical method is very striking; and algebraic methods applied to frames, such as the Warren girders, in which there are numerous similar pieces, are found to result in frequent clerical errors, owing to the cumbrous notation which is necessary, and especially owing to the necessary distinction between odd and even diagonals.*'

18. But, whilst speaking of the geometrical solution of problems relating to the science of construction, it is impossible to pass over in silence the name of Professor Culmann, the ingenious and esteemed creator of graphical statics*, for to him are due the elegant methods of that science, which, issuing from the Polytechnic School at Zurich, are now taught in technical schools throughout the world.

Numerous questions of theoretical statics, as well as many others which relate more particularly to certain branches of practical science, are solved by Professor Culmann by a simple

read before the Royal Scottish Society of Arts (Edinburgh, 1870); and the memoir *On the application of graphic methods to the determination of the efficiencies of machinery* (Transactions of the Royal Society of Edinburgh, vol. xxviii. p. 1, 1877).

* *Die graphische Statik*, Zurich, 1866. In 1875 appeared the second edition of vol. i. with rich additions. The reader is advised to read the Preface to that second German edition, also Nos. 81 and 82. Graphical Statics have been treated since in a whole series of elementary works. See UNWIN, *Wrought Iron Bridges and Roofs*, London, 1869; BOW, *Economies of construction in relation to framed structures*, London, 1873; CLARKE, *Graphic Statics*, London, 1880; EDDY, *New constructions in Graphical Statics*, in Van Nostrand's Engineering Magazine, New York, 1877–8, and American Journal of Mathematics, vol. i., Baltimore, 1878, &c., &c.

and uniform method, which reduces itself in substance to the construction of two figures, which he calls *Kräftepolygon*, and *Seilpolygon*. And although he has not considered these figures as reciprocal, in Maxwell's sense, still they are so substantially; in particular the geometrical constructions which Culmann gives in Chapter V of his work, devoted to systems of framework (*Das Fachwerk*), almost always coincide with those derived from Maxwell's own methods.

Moreover Culmann's constructions include certain cases, (which are not treated by the English geometer,) in which it is impossible to construct the reciprocal diagrams.

19. First of all, I wish to show that the *Kräftepolygon* and the *Seilpolygon* (polygon of forces and funicular polygon) of Culmann can be reduced to reciprocal diagrams.

Let there be given in a plane (which suppose always to be the orthographic plane) n forces P_1, P_2, ..., P_n in equilibrium, then by the *polygon of forces* we understand a polygon, whose sides 1, 2, ..., n are *equipollent** to the straight lines which represent the forces†.

Take in the same plane a point O, which will be called the *pole* of the polygon of forces, and join the vertices of the above polygon to that pole; denote by (rs) the ray connecting O with the vertex common to the two sides r and s. The *funicular polygon* corresponding to the pole O is a polygon, whose vertices lie in the lines of action (which we shall call **1**, **2**, ..., **n**) of the forces P_1, P_2, P_3, ..., P_n, and whose sides are respectively parallel to the rays proceeding from O‡, in such a manner that the side comprised between the lines of action of P_r and P_s, is parallel to the ray $O(rs)$, this side will be denoted by the symbol (**rs**).

The funicular polygon will be a closed one, like the polygon of forces.

20. If the lines of action of the given forces meet in the same point (Fig. 1*a*), we have two reciprocal diagrams, since evidently the two polygons will be the orthographic pro-

* *Equipollent*, that is, equal in magnitude, direction and sense, a term due to Professor Bellavitis.

† The position of the first side of that polygon is a matter of choice.

‡ The direction only of the first side of that polygon is determinate.

jections of two pyramids, having each a polyhedral angle of n faces.

If the forces are parallel, the polygon of forces is reduced to a straight line, which corresponds to the case where the base of

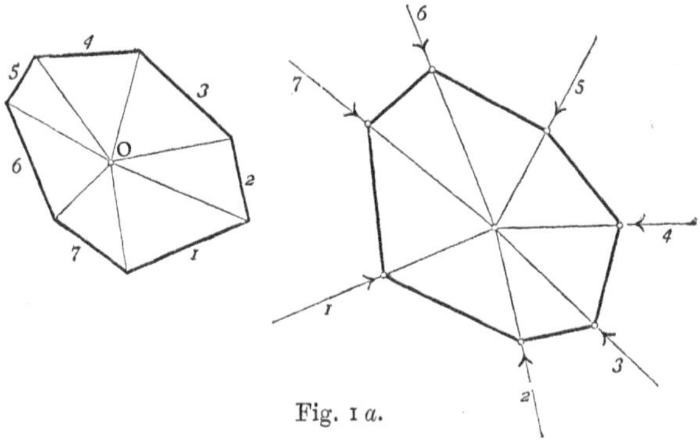

Fig. 1 *a*.

the first pyramid is perpendicular to the orthographic plane, and the vertex of the second is at infinity, that is to say, the second polyhedron is a prism having only one base at a finite distance. This case is illustrated in Figure 2*a*, in which the

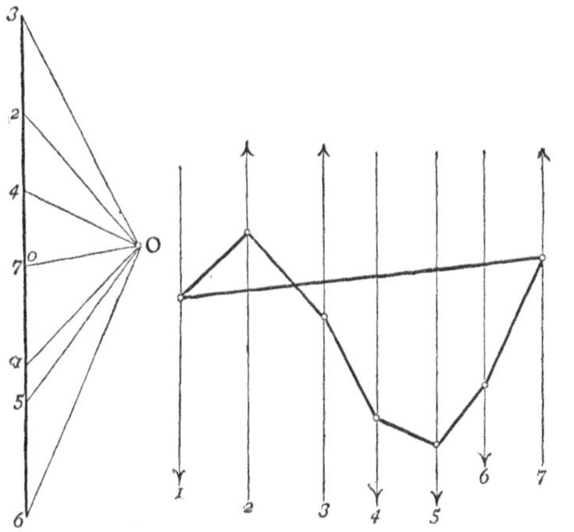

Fig. 2 *a*.

sides of the polygon of forces are not designated by one number only, but by two numbers, placed at the ends of each segment; so that the segments 01, 12, 23, 34, ..., correspond to the straight lines **1**, **2**, **3**, **4** of the second diagram.

Here, as in all which follows, we adopt in the text two series of numbers, 1, 2, 3, *r* ..., *s* ...; **1**, **2**, **3**, **r**, **s**, to distinguish the lines of the one diagram from the corresponding lines of the other.

21. Let us consider now the general case, in which the forces do not all meet in the same point.

Take a second pole O'; join it by straight lines to the vertices of the polygon of forces, and construct a second

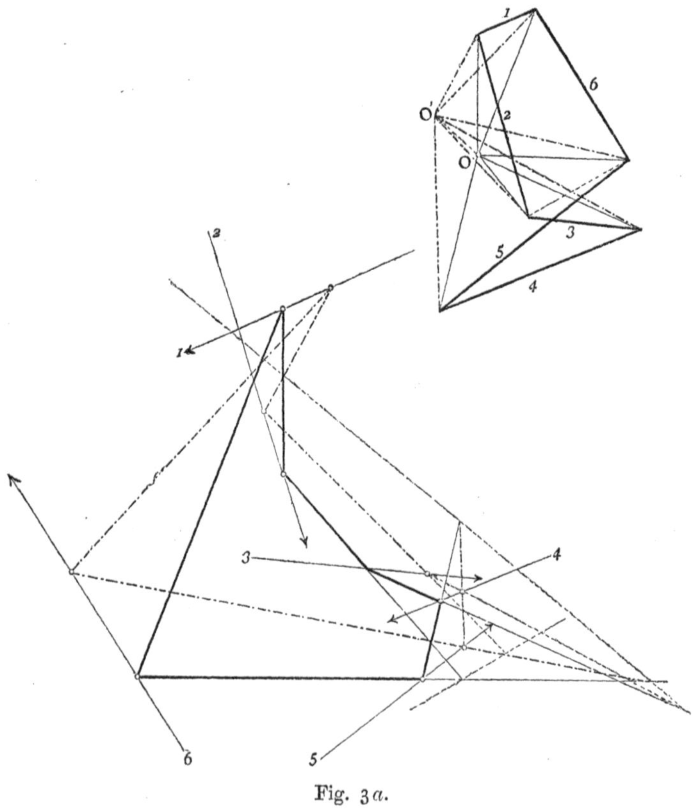

Fig. 3*a*.

funicular polygon corresponding to the new pole O', that is to say, a polygon with its sides parallel to the rays proceeding from O', and its vertices situated in the lines of action of the forces. See Figs. 3 and 5, in which the rays proceeding from the second pole O', and the corresponding sides of the funicular polygon, are denoted by dotted lines.

By operating in this way, it is plain that the two diagrams, the one formed by the polygon of forces and the rays issuing from the poles O and O', and the other formed by the two funicular polygons and the lines of action of the forces, are two reciprocal figures. The first is the projection of a polyhedron*, formed by two solid angles of n faces, whose corresponding faces form by their respective intersections a twisted polygon† of n sides; the second is the projection of a polyhedron comprised within two plane polygons of n sides, in such a way that the sides of the one meet the corresponding sides of the other. The straight line, in space, which joins the vertices of the two solid angles of n faces of the first polyhedron is conjugate to the straight line, which the two planes of the bases of the second polyhedron have in common. As a result of this, and of the property that two conjugate straight lines are orthographically projected into two parallel straight lines, it follows, that any two corresponding sides whatever (**rs**), (**rs**)$'$ of the two funicular polygons, intersect in a fixed straight line, parallel to that which joins the two poles O and O'.

This theorem is fundamental in Culmann's methods.

22. If we make the two poles O and O' coincide, the corresponding sides of the two funicular polygons are parallel (Fig. 4*a*). In this case the straight line which joins the vertices of the solid angles of the first polyhedron is perpendicular to the orthographic plane, whilst the bases of the second polyhedron are parallel.

23. The diagonal which joins the vertices of two tetrahedral angles of the first polyhedron (Art. 21), or what is the same thing, the diagonal between two vertices of the twisted polygon, is conjugate to the line of intersection of the corresponding quadrilateral faces of the second polyhedron, which

* This polyhedron has $3n$ edges, $2n$ triangular faces, 2 polyhedral angles of n faces, and n of 4 faces; the other polyhedron has $3n$ edges, $2n$ trihedral angles, 2 bases which are polygons of n sides, and n quadrilateral faces.

† If this polygon degenerates into a continuous curve, the polygon of forces, and the funicular polygon become respectively the curve of forces, and the funicular curve (catenary) of a plane continuous system of forces.

line unites the point common to two sides of the one of the bases to the point common to the corresponding two sides of the other base. In an orthographic projection, the first straight line is a diagonal joining the two vertices $(r, r+1)$, $(s, s+1)$ of the polygon of forces, that is to say, a straight line equipollent to the resultant of the forces $P_{r+1}, P_{r+2} \ldots, P_s$; the second straight line is the line of action of the same resultant. Hence the line of action of the resultant of any

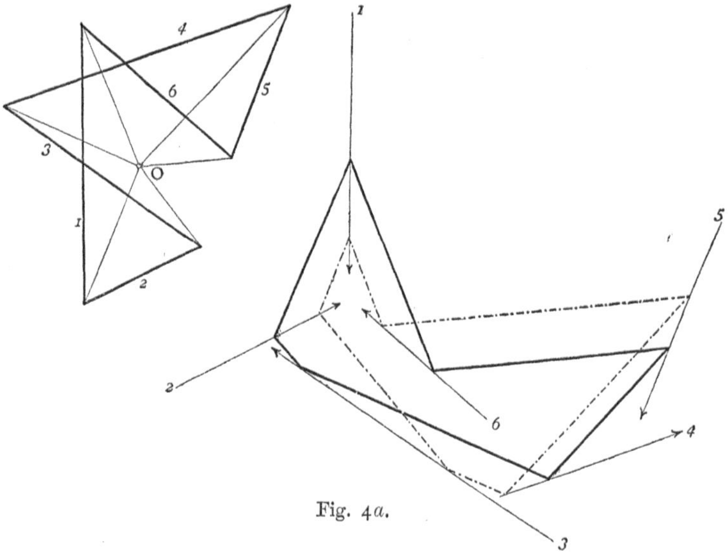

Fig. 4*a*.

number whatever of consecutive forces $P_{r+1}, P_{r+2}, \ldots, P_s$ passes through the point common to the sides $(\mathbf{r}, \mathbf{r+1})(\mathbf{s}, \mathbf{s+1})$ of the funicular polygon; another fundamental theorem of graphical statics. (See, for example, Fig. 3*a*, the resultant of the forces 6, 1, 2.)

24. If the diagonal in question of the first polyhedron is perpendicular to the orthographic plane, the conjugate straight line is at infinity. Two vertices of the polygon of forces $(r, r+1)$, $(s, s+1)$ will then coincide in one point A (see Fig. 5*a*, where $r = 1$, $s = 4$), and the sides $(\mathbf{r}, \mathbf{r+1})$, $(\mathbf{s}, \mathbf{s+1})$ of each of the funicular polygons are parallel.

The magnitude of the resultant of the forces $P_{r+1}, P_{r+2}, \ldots, P_s$ will be infinitely small, and its line of action the straight line at infinity of the orthographic plane; it is consequently an infinitely small force, acting at infinity, equivalent to a couple acting in the aforesaid parallel sides of the funicular polygon, and represented in magnitude by the straight line which joins the corresponding pole O to the point A. Since these two forces are equivalent to the system of forces P_{r+1}, P_{r+2}, ..., P_s, the one which acts along the side (**r**, **r**+**1**) is

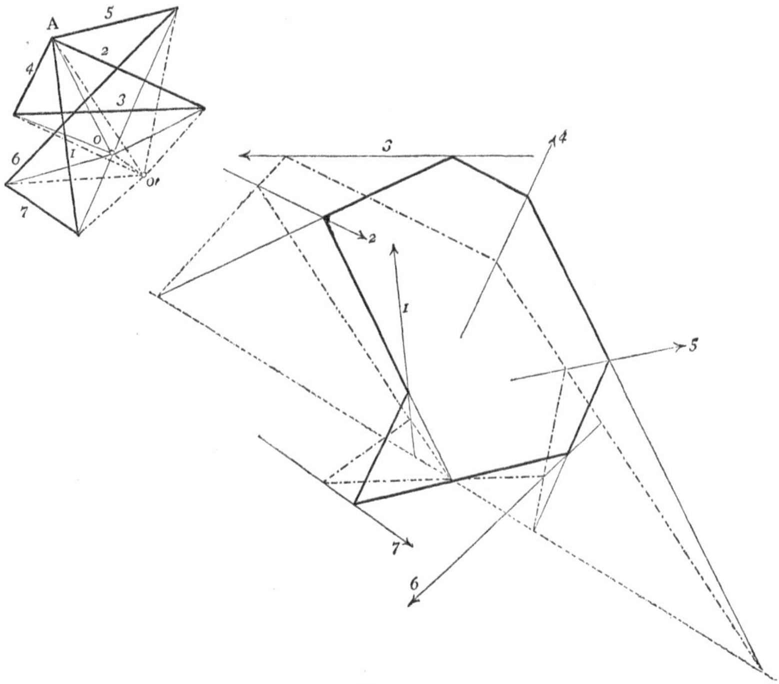

Fig. 5*a*.

directed from A towards O; and the one which acts along the side (**s**, **s**+**1**) is directed from O towards A.

25. Given the forces P_1, P_2, P_3, ..., P_{n-1} (Art. 19), the two polygons (that is the force and funicular polygons) serve to determine the force P_n, equal and opposite to the resultant of the given forces (see Fig. 3, in which $n = 6$). In fact, if we construct a crooked line 1, 2, 3, ..., $(n-1)$, whose sides are equipollent to the given forces; it is clear that the straight line n which joins the extremities of the crooked line (when its direction is from the final to the initial point) is equipollent to P_n. Next take a pole O, and construct a funicular polygon, whose first $n-1$ vertices **1, 2, 3**, ..., **(n−1)**, lie in the lines of action of the given forces $P_1, P_2, P_3, \ldots, P_{n-1}$; and whose sides **(n, 1) (1, 2) (2, 3)** ... **(n−1, n)** are respectively parallel to the rays connecting O with the similarly named vertices of the first polygon. Then the straight line drawn through the last vertex **n** of the funicular polygon, (that is to say through the point where the first side **(n.1)** meets the last **(n−1, n)**,) parallel to the last side n of the polygon of forces, is the line of action of P_n.

If the first side of the funicular polygon passes through a fixed point, and the pole O moves in a straight line, then all the sides pass through fixed points situated on a straight line parallel to the one described by the pole O (Art. 21). This is contained in the celebrated porism of Pappus:

'*Si quotcumque rectae lineae sese mutuo secent, non plures quam duae per idem punctum, omnia autem in una ipsarum data sint, et reliquorum multitudinem habentium triangulum numerum, hujus latus singula habet puncta tangentia rectam lineam positione datam, quorum trium non ad angulum existens trianguli spatii unumquodque reliquum punctum rectam lineam positione datam tanget**.'

26. If we consider the point O to be capable of occupying any position whatever in the plane, the properties of the two polygons (that is the polygon of forces and the funicular polygon) may be compendiously stated in the following geometrical enunciation:

Let a plane polygon be given of n sides $1, 2, 3, \ldots, (n-1), n$; and, in the same plane, $n-1$ straight lines **1, 2, 3**, ..., **(n−1)**, respectively parallel to the first $n-1$ sides of the polygon. Join the point O (i.e. a pole, moveable in any manner whatever in the plane) to the vertices of the given polygon. Imagine further a variable polygon of n sides, the first $n-1$ vertices of which **1, 2, 3**, ..., **(n−1)**, lie in the corresponding similarly named straight lines, whilst its n sides **(n.1)**, **(1.2)**, **(2.3)** ..., **(n−1, n)** are parallel to the rays which join the similarly named vertices of the given polygon to the pole O. Then the intersection of any two sides whatever **(r, r+1)**,

* [*Mathematicae Collectiones*, preface to Book VII. p. 162, of the edition of COMMANDINO (Venice, 1689). See also the translation or paraphrase of the porism, given by PONCELET in No. 498 of his *Traité des propriétés projectives* (Paris, 1822)].

$(s.s+1)$, of the variable polygon lies on a determinate straight line, parallel to the diagonal which joins the vertices $(r.r+1)$, $(s, s+1)$ of the given polygon.

This theorem, which is not very readily proved by means of the resources of *Plane Geometry* alone, is on the contrary self-evident, if we consider the two plane figures as orthographic projections of two reciprocal polyhedra.

27. The polygon of forces is the projection of a plane polygon, or twisted polygon, according as the directions of the forces P do or do not meet in the same point. As we have seen in Art. 20, one of the two reciprocal diagrams in the first case is formed by the polygon of forces and the pole O, the other, by the lines of action of the forces, and the funicular polygon corresponding to the pole O. In the second case, on the contrary, another pole O' must be added to the first diagram, and to the second a funicular polygon corresponding to this pole O'; we have further seen from Art. 22 that the two poles may be made to coincide, and that then the first diagram becomes as simple as possible. But, if we wish on the other hand to simplify the second, it is best to remove the pole O' to infinity in an arbitrary direction; and then the polyhedron, of which the first diagram is the orthographic projection, has the vertex of one of its polyhedral angles at infinity; and since the polar plane of a point at infinity is parallel to the central axis, the new funicular polygon corresponding to O' has all its sides on the same straight line (whose point at infinity is O'). The absolute position of this straight line in the orthographic plane is still arbitrary, and therefore it may be removed to infinity.

Very simple results are also obtained by the following method:

Suppose that the previously mentioned polyhedral angle of the first solid coincides with the infinite point of the central axis; in the first diagram the pole O alone appears, since the edges corresponding to the other polyhedral angle are projected orthographically into the vertices of the polygon of forces. The polar plane of the vertex O' is now at infinity; hence the whole of the second funicular polygon is at an infinite distance (see Art. 13).

28. We conclude from these very simple cases, that it is possible to consider the polygon of forces, and the funicular polygon, of a system of forces in equilibrium, situated in a plane (the *orthographic plane*), but not meeting in the same point, as reciprocal diagrams. The one diagram is formed by the polygon of forces and the rays joining its vertices to a pole O, and the other by the lines of action of the forces, the funicular polygon relative to the pole O, and the straight line at infinity; the first diagram is simply the projection of a polyhedron, whose faces are obtained by projecting the n sides of a twisted polygon perpendicularly to the orthographic plane, from a point in space at infinity. The reciprocal polyhedron, which has for its projection the second diagram, is the infinite portion of space, limited by a plane polygon and the n planes passing through the sides of that polygon and prolonged everywhere to the plane at infinity.

CHAPTER III.

APPLICATION OF RECIPROCAL DIAGRAMS TO FRAMEWORK.

29. LET us pass on now to the study of the more complicated diagrams, which present themselves in the theory of frames*. Consider two polyhedral reciprocal surfaces Σ and Σ', which possess an 'edge,' are simply connected†, and whose edges are two closed twisted polygons‡; let Π be the polyhedron enclosed by the surface Σ, and the pyramidal surface whose vertex is a point Ω, taken arbitrarily in space, and whose base-line is the polygonal edge of Σ; let Π' be the polyhedron reciprocal to Π, i.e. the polyhedron enclosed by the surface Σ', the polar plane ω of Ω, and the planes of the angles of the polygonal edge of Σ'. Project orthographically the two polyhedra, and we obtain two reciprocal diagrams, which we will now proceed to study.

Suppose that the polygonal edge of Σ has n sides, and that the surface has besides these m ordinary edges§, and p faces. The polyhedron Π will have $n+p$ faces, and $2n+m$ edges, and therefore $m+n-p+2$ vertices. Hence Σ has, besides those on its polygonal edge, $m-p+1$ vertices‖.

* A Frame is a structure composed of bars or rods attached together by joints, which are considered merely as hinges or pivots. Let AB be any one bar (whose weight is neglected) of such a frame; and assume that no force acts upon it, except at the joints A, B. Then the whole of the forces (some external, some consisting of pressures from the bar or the bars which meet it at the joint A) acting on it at the joint A can be reduced to a single resultant: so may those at the joint B; and these resultants being necessarily equal and opposite, must act along the bar AB. Hence the bar is in a simple state of *tension*, when these resultants act outwards; or of *compression*, or *thrust*, when they act inwards. A bar is called a *tie* when in tension; a *strut* when in compression (CROFTON, *Lectures on Applied Mechanics, at the Royal Military Academy*, London, 1877).

† A surface with an edge is *simply connected*, if its edge is a single closed continuous line which does not intersect itself.

‡ If the edge of Σ is a plane polygon of n sides, that of Σ' will be a point, the vertex of a polyhedral angle of n faces.

§ We have evidently $m \geqq n$.

‖ Therefore m can never be less than $p-1$.

Reciprocally, Π' has $m+n-p+2$ faces, $n+p$ vertices, and $2n+m$ edges.

30. Suppose now that the projection of Σ' is the skeleton of a frame with p joints, and m rectilinear bars, and that the external forces which are applied to it have for their lines of action the projections of the sides of the polygonal edge of Σ', and are represented in magnitude by the n sides of the projection of the polygonal edge of Σ*. Then the projection of the face of Π', which lies in the plane ω, will be the funicular polygon of the external forces, corresponding to the pole O, the projection of Ω; and the projections of the m edges of Σ, not pertaining to its polygonal edge, represent the values of the internal forces or stresses to which the corresponding bars of the structure are subjected, in consequence of the given system of external forces.

31. If the point Ω is removed to infinity in a direction perpendicular to the orthographic plane, the plane ω will coincide with the plane at infinity. Then the first diagram reduces to the projection of Σ, i.e. to the entire system of the straight lines which represent the magnitudes of the external and internal forces; and the second diagram, from which the funicular polygon has completely disappeared, merely contains the skeleton of the structure (i.e. the lines of actions of the internal forces), and the lines of action of the external forces. In the figures which accompany the text, the first diagram is indicated by the letter b, and the second by the letter a.

32. If the external forces are all parallel to one another, as very frequently happens in practice, the edge of Σ will be a polygon situated entirely within a plane perpendicular to the orthographic plane; and therefore the sides of the polygon of external forces will all fall on one and the same straight line.

33. The diagrams may be formed by other degenerate polygonal figures arising from analogous degenerations of the figures in space.

Suppose, for example, that we have in space a solid tetrahedral angle, corresponding to a quadrilateral face in the

* This is only possible when Σ has no vertex at infinity; i.e. when Σ' has no face perpendicular to the orthographic plane.

reciprocal figure; and let two edges (not opposite) of the solid angle approach one another indefinitely, in their plane, and ultimately coincide. The solid tetrahedral angle will be replaced by a system composed of a trihedral angle and a plane passing through one of its edges. Consequently the quadrilateral face of the reciprocal figure will have two sides which, without ceasing to have a common vertex, will be superposed and may have either the same or the contrary direction.

Passing from the figures in space to their orthographic projections, we shall have in one of the diagrams a point from which four straight lines diverge, two of which will be superposed; and in the other diagram a quadrilateral with three collinear vertices*.

34. Given the skeleton of a framework and the system of external forces, it is necessary first of all to construct the polygon of these forces, i.e. a polygon whose sides are equipollent to them. In the figures contained in this work both the external forces and the sides of their polygon are denoted by the numbers 1, 2, 3,..., so disposed that, if we go round the contour of the polygon in the increasing order of the numbers, each side is passed over in the sense of the force which it represents. This way of going round the polygon is called the *cyclical order* of its contour.

When we wish to construct the diagram reciprocal to the one formed by the bars of the frame and by the lines of action of the external forces, the order in which the forces are made to follow one another when their polygon is constructed is not arbitrary; this order is determined by the following considerations:

In the polygon of external forces, which forms part of the diagram b, the sides equipollent to two forces will be adjacent, when the lines of action of those forces belong to the contour of the same polygon in diagram a, because that polygon corresponds to the vertex which is common to those two sides.

Let us then give the index 1 to any one whatever of the external forces; the line of action of the selected force is a side common to two polygons of diagram a; the contour of

* Examples of these degenerate forms are to be found at p. 444 and in the first two tables of the memoir of Professor Fleeming Jenkin, already cited on p. 132, 1869, and in Fig. 9 of our examples.

each of these contains the line of action of another external force; thus there are two external forces which may be regarded as contiguous to the force 1, and the index 2 may be attributed to either of them indifferently, and the index n to the other, where n is the number of external forces. After this, the order of the other sides of the polygon of external forces is completely determined. Suppose that the joints, to which the external forces are applied, all lie on the contour* of the skeleton of the framework, then the forces must be taken in the order in which we meet them in passing round the contour. When we do not follow these rules, as well as those previously laid down, we are still able to determine the internal forces graphically, but we no longer have two reciprocal diagrams, and the figures will be very complicated; since any segment which does not lie in its proper place will have to be repeated or removed to another place in view of further† constructions; just what happens in the old method, which consists in constructing a polygon of forces for each separate joint of the framework.

35. The polygon of external forces being thus constructed, we complete the diagram *b*, by constructing successively the polygons which correspond to the different joints of the framework.

The problem, of constructing a polygon all of whose sides have given directions, is soluble when only two of the sides are unknown. For this reason we ought to commence at a joint through which only three straight lines pass; the lines of resistance of two bars, and the line of action of an external force. The segment equipollent to the external force will be a side of the triangle corresponding to the joint in question, and consequently we are able to construct the triangle.

* The contour of certain structures (trusses) is composed of two systems of bars, an upper and lower. The bars which unite the joints of one of these systems to those of the other (we consider them as going from the upper to the lower) are *diagonals* or *braces*, if they are inclined from left to right, and if inclined in the opposite sense *contra-diagonals*. We call the upright bars *verticals*.

† For this reason Figs. 1 and 3 of Pl. xvi. in the atlas of Culmann's *Graphical Statics* are not reciprocal, and similarly Figs. 7 and 7, of pl. xix., &c.; on the other hand, diagrams 168 and 169 of p. 422 (1st edition) are perfectly reciprocal.

The construction presents no ambiguity, if we remember that to a bar of the framework belonging to the contour of a polygon of the diagram *a*, to which the lines of action of two external forces also belong, corresponds in the diagram *b* a straight line passing through the vertex common to the sides equipollent to those two forces.

Then we pass on successively to the other joints, taking them in such an order that in each new polygon to be constructed only two unknown sides remain.

In the figures given, all the lines of each of the diagrams have numbers attached to them indicating in what order the operations are to be performed.

'*The figure can be drawn in a few minutes, whereas the algebraic computation of the stresses, though offering no mathematical difficulty, is singularly apt, from mere complexity of notation, to result in error**.'

36. A superficial consideration might lead us to conclude that the solution of the above problem is possible and determinate, even in the case where the frame has no joint at which three straight lines only intersect†.

Suppose, for example, that the skeleton of the structure is formed by the sides **5, 6, 7, 8** of a quadrilateral and the straight lines **9, 10, 11, 12** which join its vertices to a fifth point; and let the external forces **1, 2, 3, 4** be applied at the vertices (**8, 5, 9**), (**5, 6, 10**), (**6, 7, 11**), (**7, 8, 12**) of the quadrilateral‡. Construct the polygon 1, 2, 3, 4 of external forces and through the points (1, 2), (2, 3), (3, 4), (4, 1) respectively, draw the indefinite straight lines 5, 6, 7, 8.

Then our problem is, to construct a quadrilateral, whose sides 9, 10, 11, 12 are respectively parallel to the lines denoted by these numbers in the given diagram, and whose vertices (9, 10), (10, 11), (11, 12), (12, 9) lie respectively on the straight lines 5, 6, 7, 8. Since the problem of constructing a quadrilateral whose sides have given directions (or pass through given

* Professor Fleeming Jenkin, p. 443 of the volume of the Transactions of Edinburgh already cited on p. 132.

† The frame or truss is always supposed to be formed by triangles only.

‡ Exactly the same reasoning applies to the structure formed by any polygon whatever, and the straight lines joining its vertices to a fixed point.

We have given no figures for this article, but the reader can easily supply them for himself.

points on the same straight line), and whose vertices lie on four fixed straight lines admits in general of one, and only one, solution; we might at first sight suppose that the diagram of forces is completely determined.

But this illusion vanishes when we remember that the geometrical problem presents certain cases which are impossible and indeterminate. In a word, suppress one of the conditions, that is, assume that the quadrilateral has its sides parallel to the given directions, and that its first three vertices only lie on given straight lines 5, 6, 7; then we know that the fourth vertex describes a straight line r* whose point of intersection with the given straight line 8, will determine the fourth vertex, and give the required solution. Now if the data are such that r is parallel to 8, we arrive at an impossibility. Again, making a still more special hypothesis, if the straight line r coincides with 8, the problem is indeterminate, and an infinite number of quadrilaterals will satisfy the conditions of the problem.

In order to show that the construction of the diagram reciprocal to the given diagram is indeterminate or impossible, it is enough to reflect, that if we consider the given diagram as the polygon of forces whose magnitudes are represented by the segments **5**, **6**, **7**, **8**, the pole of the polygon being the point (**9**, **10**, **11**, **12**), then the reciprocal diagram (9, 10, 11, 12) is simply the corresponding funicular polygon. But, in order that the construction of the funicular polygon may be possible, it is necessary that the forces should be in equilibrium: if then we suppose the magnitude of the forces **5**, **6**, **7**, **8** given, and also the lines of action 5, 6, 7 of three of them, the line of action of the fourth is perfectly determined, and is the straight line r of which we have just spoken. Hence if r and 8 do not coincide, the forces in question **5**, **6**, **7**, **8** are not in equilibrium, but are equivalent to an infinitely small force at an infinite distance, and the problem is impossible; if, however, r and 8 do coincide, that is to say, if the forces in question are in equilibrium, the problem is indeterminate, since for a given pole and system of forces we are able to construct an infinite number of funicular polygons.

* This is the Porism of PAPPUS, for which see CREMONA's *Projective Geometry*, Art. 114.

In the first of these two cases, equilibrium might be obtained by combining the forces **5**, **6**, **7**, **8** with a force equal and opposite to their infinitely small resultant, situated at infinity, i.e. by considering the polygon **5**, **6**, **7**, **8** as the projection not of a quadrangle but of a pentagon, two successive vertices of which project into one and the same point (**7**, **8**, **12**). The straight line **12** would then be the projection of two distinct straight lines in space, and consequently in the reciprocal diagram, to the point (**9**, **10**, **11**, **12**) there would correspond an open pentagon 9, 10, 11, 12, 12′, having its vertices (9, 10), (10, 11), (11, 12), (12′·9) situated respectively on the straight lines 5, 6, 7, 8, and its vertex 12, 12′ at infinity.

37. Each rectilinear bar of a framework is the line of action of two equal and opposite forces, applied respectively at the two joints connected by the bar. The common magnitude of these two forces, that is to say, the measure of the stress which they exert on the bar, is given by the length of the corresponding straight line of diagram *b*. These two forces may either be considered as actions or as reactions: to pass from one case to the other, it is only necessary to reverse their directions*.

38. Each joint of the framework is the point of application of a system of at least three forces, in equilibrium; one of them may be an external force, the others are the reactions which are called into play in the bars which meet at the joint in question. It is sufficient to know the sense of one of these forces in order to obtain that of all the others. Two cases are possible.

First, if an external force be applied at the joint considered; then if we pass along the corresponding side of the polygon of forces in the sense of that force, each of the other sides of the polygon will be passed over in the sense which belongs to its corresponding internal force, considered as a reaction applied at the joint in question. If, on the contrary, we wish to find the sense in which the internal forces would act when considered as actions, it is sufficient to reverse the direction of the external force.

* In the figures of this work the ties are shown by finer lines than the struts. In the figures of Culmann and Reuleaux the struts are shown in double lines, the ties by single ones. See the first note on p. 143.

If the only forces which act at the joint in question are internal forces, it is likewise sufficient to know the sense of one of them in order to find by the process just explained the sense of all the others.

We shall call that order which corresponds to the internal forces considered as actions the *cyclical order* of the *contour* of a polygon of the diagram *b*. We see then, that by commencing at any joint at which an external force is applied, we are able to determine in succession the magnitude and sense of all the internal forces. By considering one of the internal forces as an action applied at one of the two joints between which it acts, we are able to recognise at once whether the bar connecting the same joints is in compression or tension.

Every straight line in diagram *b* is a side common to two polygons: in going round the contour of each of them in their respective cyclical order, the sides will be described once in the one sense, and once more in the contrary sense *.

This corresponds to the fact that the straight line in question represents two equal and opposite forces acting along the corresponding bar of the framework.

39. We know that the algebraic sum of the projections of the faces of a polyhedron is equal to zero. By applying this theorem to the polyhedron Π (Art. 29), remembering that the projection of the surface Σ forms the polygons of the diagram *b*, corresponding to the joints of the structure, whilst the projection of the rest of the polyhedron Π is simply the polygon of external forces, we arrive at the following theorem:

Regarding the area of a polygon as positive or negative according as that area lies to the right or to the left of an observer passing round its contour in the cyclical order which belongs to it, then the sum of the areas of the polygons of diagram b which correspond to the joints of the framework is equal and opposite to the area of the polygon of external forces.

Clerk Maxwell has arrived at this theorem in another

* This property is in accordance with the so-called *Law of Edges* (KANTENGESETZ) of polyhedra possessing one internal surface and one external.

See MÖBIUS, *Ueber die Bestimmung des Inhalts eines Polyeders* (Leipziges Berichte, 1865, vol. 17, p. 33 and following), or *Gesammelte Werke*, 2nd Band, p. 473; also BALTZER, *Stereometrie*, § 8, Art. 16.

way by investigating whether it is possible or not to construct the diagrams of forces * for any plane frame whatever.

40. The *method of sections* generally employed in the study of variable systems furnishes a valuable means of verification.

If an ideal section be made in the structure, then in each of the parts so obtained, the external forces are in equilibrium with the reactions of the bars cut across by the section.

If only three of the reactions are unknown, we can deduce them from the conditions of equilibrium, since the problem of decomposing a force P into three components, whose lines of action **1**, **2**, **3** are given and form with 0, the line of action of P, a complete plane quadrilateral, is a determinate problem and admits of only one solution.

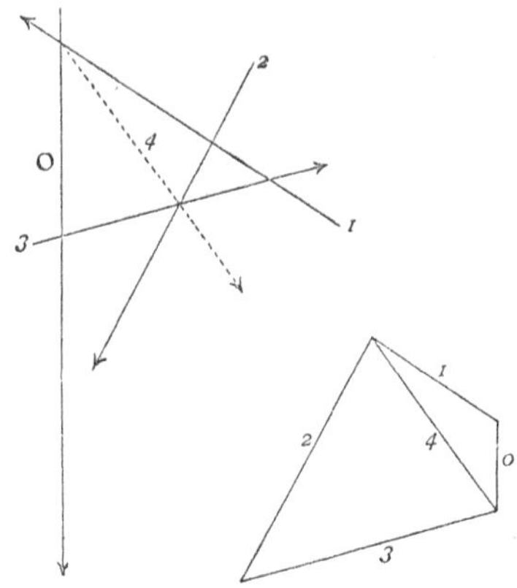

Fig. 6.

In fact (Fig. 6) it is only necessary to draw one of the diagonals of the quadrilateral, for example, the straight line **4** which joins the points (**0**, **1**), (**2**, **3**); to decompose the given force 0 into two components along the straight lines **1**, **4** (we do this by constructing the triangle of forces 0, 4, 1 of which the side *o* is given in magnitude and direction); and finally to decompose the force **4** along the straight lines **2** and **3** (by constructing in like manner the triangle of forces 4, 3, 2).

This method, which may be called the *static method*, is all-sufficient for the graphical determination of the internal forces, equally with the geometrical method, previously explained, which is deduced from the theory of reciprocal figures, consists in the successive construction of the polygons corresponding to the different joints of the structure. The static method is at least as simple, it can be rendered very useful in combination with the latter method, and it permits the rapid verification of the constructions. The external forces applied to a portion of the structure, obtained by means of any section

* Memoir of 1870, p. 30, already cited on p. 132.

whatever, and the reactions of the bars that are cut, must have the property that the corresponding lines of diagram *b* form a closed polygon. This polygon must be the projection of a closed twisted polygon, and not merely of an open crooked line whose extremities are situated in a straight line perpendicular to the orthographic plane; this condition requires that the twisted reciprocal polygon shall also be closed, i.e. we are able to unite the corresponding lines of the diagram *a* by a closed funicular polygon.

The method of sections may also be presented in another form. Denoting again by 0 the resultant of all the known forces applied to the portion of the structure considered, and by **1**, **2**, **3** the three unknown reactions, the sum of the moments of these four forces in regard to any point whatever is zero. Now, by taking as the centre of moments the point where two lines of resistance meet, for example, the point (**2**, **3**), the moment of the third reaction **1** will be equal and opposite to that of the force 0. We thus obtain a proportion between four magnitudes (the two forces and their moments) among which the only unknown quantity is the magnitude of the force **1**. This is the method of *statical moments*, by which the internal forces developed in the different bars of a framework can be calculated numerically, instead of being constructed graphically *.

* See A. Ritter, *Elementäre Theorie und Berechnung eiserner Dach- und Brücken-Constructionem*, 2nd edition (Hannover, 1870).

CHAPTER IV.

EXAMPLES OF FRAME- AND STRESS- DIAGRAMS.

41. WE will now pass on to the study of some suitable examples to show the simplicity and elegance of the graphic method. We do not always adhere to regularity and symmetry of form in the structures which we are about to study, although in practice engineers hardly ever depart from these conditions. *But the symmetrical forms of practice are only particular cases of the irregular ones of abstract geometry*: and therefore the forms which we shall treat include all the cases which are possible in practice. In what follows, the expression 'framed structure' will be used in the general and theoretical sense which Maxwell attributed to the word frame.

'*A frame is a system of lines connecting a number of points. A stiff frame is one in which the distance between any two points cannot be altered without altering the length of one or more of the connecting lines of the frame.—A frame of s points in a plane requires in general* $2s-3$ *connecting lines to render it stiff* *.'

We confine ourselves to the study of plane figures formed by triangular parts.

42. As a first (general and theoretical) example, let **1**, **2**, **3**, ..., **10** (Fig. 7 *a*) be a system of ten external forces in equilibrium; construct the corresponding polygon of forces, and join its vertices to an arbitrary pole *O* (Fig. 7 *b*, in which the polygon of forces is represented by double lines). Draw next a funicular polygon, having its sides respectively parallel to the rays from *O* (Fig. *b*), and its vertices lying in the lines of action of the forces **1**, **2**, **3**, ..., **10**. The forces in question are applied at the different joints of a framed structure, the bars of which are numbered from **11** up to **27** (Fig. 7 *a*).

* Page 294, Phil. Mag., April 1864.

We commence by constructing the triangle corresponding to the joint (**10**, **11**, **12**), drawing through the extremities of the straight line 10 two straight lines 11, 12, respectively

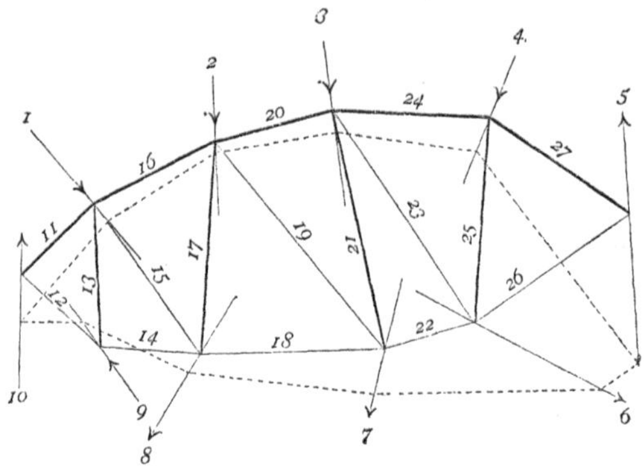

Fig. 7*a*.

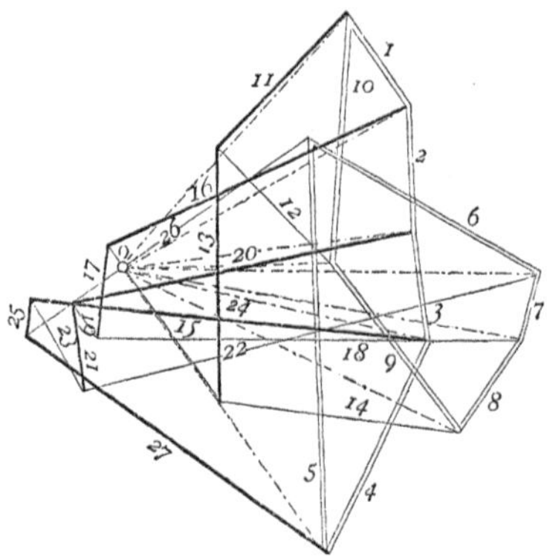

Fig. 7*b*.

parallel to **11** and **12**; we notice that the straight line 11 must pass through the point (1, 10), because in the diagram *a*, the lines **1**, **10**, **11** belong to the contour of the same polygon*; for the same reason 12 must pass through the point (9, 10). Passing round the contour of the triangle just

* This polygon is a quadrilateral, whose fourth side is the side of the funicular polygon comprised between the forces 1 and 10. As previously stated (Arts. 27, 31) the whole of the funicular polygon might have been removed to infinity.

obtained, in a sense contrary to that of the force **10**, we obtain the sense of the actions called into play at the joint we are dealing with, along the lines **11** and **12**; and it is thus seen that the bar **11** is in a state of compression and the bar **12** of tension.

Now construct the quadrilateral corresponding to the joint at which the force **9** is applied, and for this purpose, draw 13 through the point (11, 12) and 14 through the point (8, 9). The bar **13** is in compression, **14** in tension.

Next construct the pentagon corresponding to the joint at which the force **1** is applied, by drawing 15 through the point (13, 14), and 16 through the point (1, 2). The pentagon thus obtained is a crossed one. The bar **15** is in tension, **16** in compression.

Then construct the pentagon corresponding to the joint at which the external force **8** is applied; by drawing the line 17 through the point (15, 16), and the line 18 through the point (7, 8). The bar **17** is in compression, **18** in tension.

Continuing in this manner we find all the other internal forces. The last partial construction gives the triangle which corresponds to the point of application of the force **5**. The bars **20**, **21**, **24**, **25**, **27** are in compression; **19**, **22**, **23**, **26** are in tension.

43. Figure 8*a* represents a bridge girder, at the joints of which are applied the forces **1**, **2**, **3**, ..., **8**, **9**, **10**, ..., **16** all vertical; the forces **1** and **9** acting upwards represent the reactions of the supports; the forces **2**, **3**, ..., **8** are the weights applied at the joints on the upper platform; and **10**, **11**, ..., **16** the weights applied at the joints of the lower platform.

These forces are taken in the order in which they are met with in going round the contour of the structure; and in diagram *b* the sides of the polygon of external forces are disposed in the same order. This polygon has all its sides lying in the same vertical straight line; the sum of the segments 1, 9 is equal and opposite to that of the segments 2, 3, ... 8, 10, 11, ..., 16, because the system of external forces is necessarily in equilibrium.

The diagram *b* is completed by following precisely the same rules as those just laid down. Commence at the joint (**1**, **17**, **18**); draw the straight line 17 through the point (1, 2), where the upper extremity of the segment 1 meets the upper extremity of the segment 2 ; and the straight line 18 through the point (16, 1), which is both the lower extremity of the segment 16 and that of the segment 1.

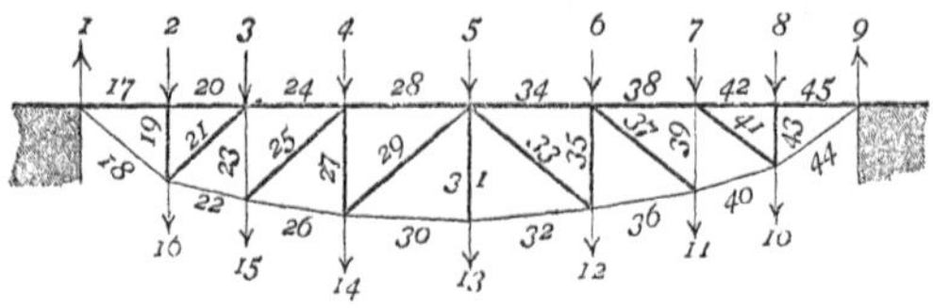

Fig. 8*a*.

Pass on to the joint (**2, 17, 19, 20**). Draw 19 through the point (17, 18), and 20 through the point (2, 3), the lower end of 2 and upper extremity of 3; and we obtain the polygon 2, 17, 19, 20, which is a rectangle.

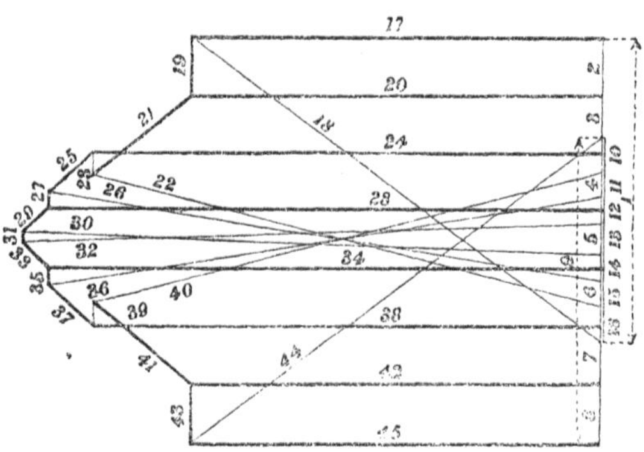

Fig. 8*b*.

Construct the polygon corresponding to the joint (**16, 18, 19, 21, 22**). For this purpose draw 21 through the point (19, 20), and 22 through the point (15, 16); we thus obtain a crossed pentagon. Continue to deal in the same manner with each of the points of application of the forces **3, 15, 4, 14, 13, 5, 12, 6, 11, 7, 10, 9**, taken in succession.

Since the diagram *a*, which represents the skeleton of the structure and all the external forces, has for its axis of symmetry the vertical which passes through the centre of the figure, the diagram *b* has for its axis of symmetry the median horizontal line. For example, the triangle 9, 45, 44 is symmetrical to the triangle 1, 17, 18; the rectangle 8, 45, 43, 42 to the rectangle 2, 17, 19, 20: and so on.

All the upper bars are in compression, and all the lower ones are in tension.

The diagonals and contra-diagonals are all in compression; finally two of the verticals **23**, **39** are in tension, and all the rest in compression.

44. Figure 9*a** represents one half of a locomotive shed. The external forces are the weights **1**, **2**, **3**, **4**, **5** applied at

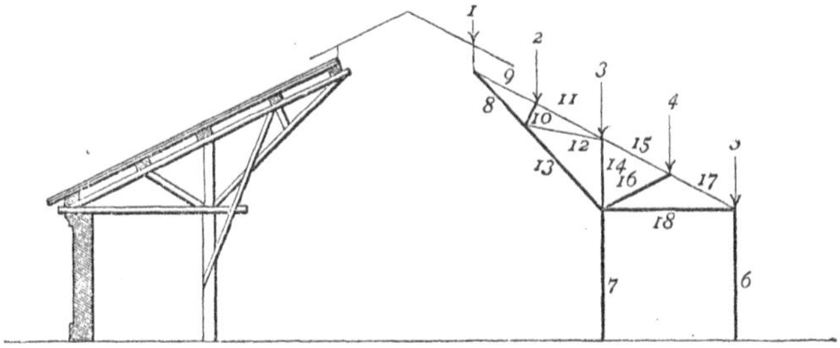

Fig. 9*a*.

the upper joints of the frame, and the reactions **6** and **7** of the wall and column. Again, all the external forces are parallel, and consequently the polygon of forces reduces, in diagram *b*, to one straight line. The force **6** (taken in the opposite sense to that in which it really acts) is equal to a certain part of the weight **5**; by adding the difference to the other weights we get the magnitude of the force **7**.

In the diagram *b* the direction of the lines 8 and 13 coincide; the first is a part of the second. Here then we have for the polygon corresponding to the joint (**8**, **10**, **12**, **13**) one of those degenerate forms about which we spoke in Art. 33; the polygon is in fact a quadrilateral 8, 10, 12, 13, having three of its vertices (13, 8), (8, 10), (12, 13) in one straight line.

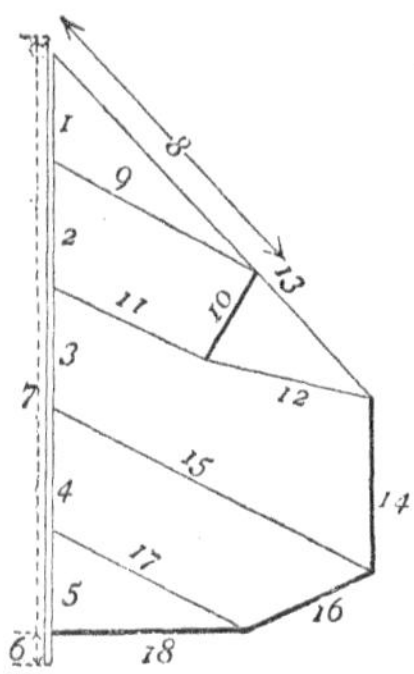

Fig. 9*b*.

The polygon 5, 17, 18, 6, corresponding to the point where

* This example is taken from Pl. xix. of the atlas of *Graphische Statik* of CULMANN, 1st edition. As previously stated, the two diagrams are not rigorously reciprocal.

the roof is supported by the wall, presents an analogous degenerate form, since the vertices (6, 5), (upper point of the segment (6), (5, 17), and (18, 6) all lie in the same straight line.

The lower bars **8**, **13**, **18** are in compression, as well as the diagonals **10**, **14**, **16**, the column **7** and the wall **6**; while the upper pieces **9**, **11**, **15**, **17** and the diagonal **12** are in tension.

45. Diagram *a* of Fig. 10 represents a truss at the upper joints of which are applied the oblique forces **1**, **2**, ..., **7**, which

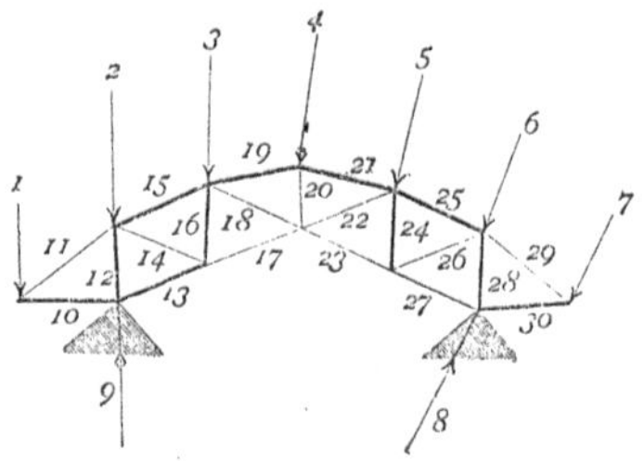

Fig. 10*a*.

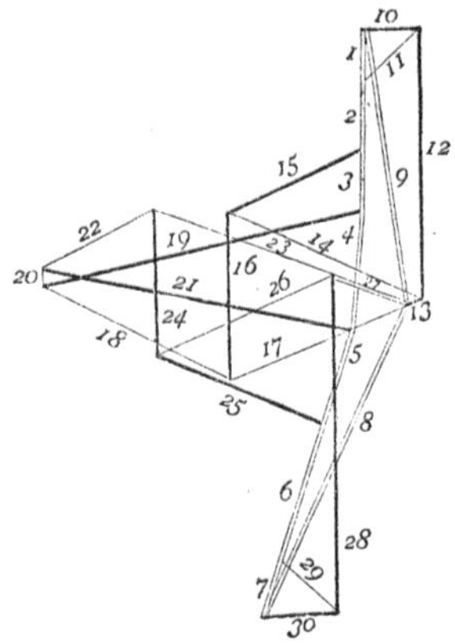

Fig. 10*b*.

may be considered as the resultants of the dead-loads and wind pressure; the forces **8**, **9** represent the reactions at the supports.

The polygon of external forces is drawn in diagram *b* with double lines.

We construct successively the triangle 1, 10, 11, the quadrilateral 9, 10, 12, 13, the pentagon 2, 11, 12, 14, 15, the quadrilateral 13, 14, 16, 17, the crossed pentagon 3, 15, 16, 18, 19; the crossed quadrilateral 4, 19, 20, 21, the pentagon 17, 18, 20, 22, 23, and so on.

The upper bars **15**, **19**, **21**, **25** are in compression, as well as the lower bars **10**, **13**, **30**, and the verticals **12**, **16**, **24**, **28**; whilst all the remaining bars of the structure are in tension.

46. The diagram *a* of Fig. 11 represents a suspension bridge, loaded at each of its upper joints with weights **1**, **2**, ..., **8**, and at each of its lower joints with weights **10**, **11**, **21**, ..., **16**; the weights are kept in equilibrium by the two oblique reactions **9**, **17** at the two extreme points of the structure*.

The polygon of external forces has its first eight sides in succession along the same vertical straight line, and its seven

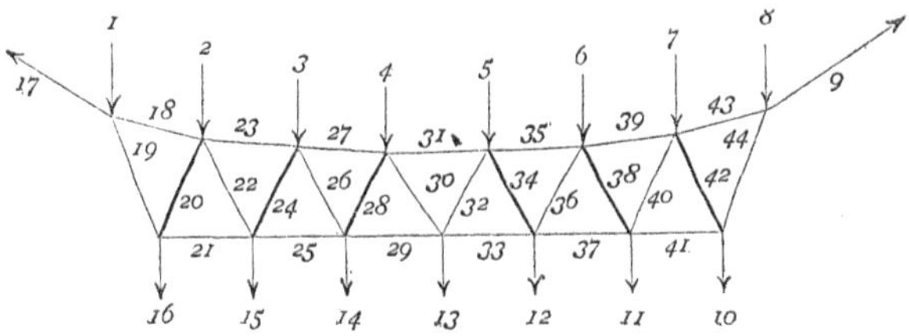

Fig. 11 *a*.

last sides situated in another vertical straight line. The oblique sides 9 and 17 intersect, so that the polygon is a crossed one. We construct successively the polygons 1, 17, 19, 18; 16, 19, 20, 21; 2, 18, 20, 22, 23; 15, 21, 22, 24, 25; 3, 23, 24, 26, 27; and so on; most of which are crossed.

Diagram *b* shows that the upper bars are all in tension, and that the tension decreases from the ends towards the

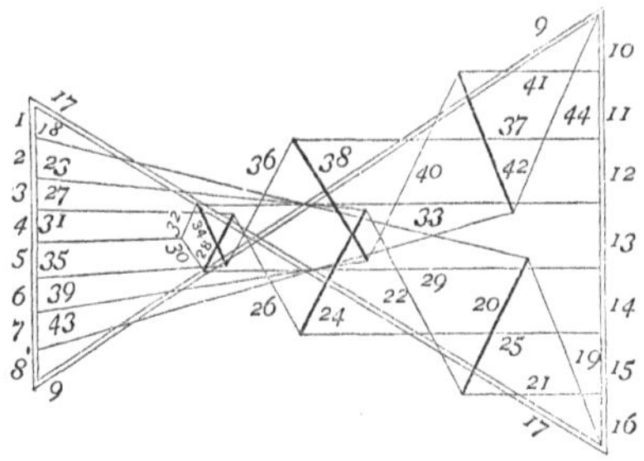

Fig. 11 *b*.

middle of the structure; the bars of the lower boom are also all in tension, but in them the tension decreases from the middle towards the ends.

The extreme diagonals and contra-diagonals are in tension; in the portion situated to the left of the axis of symmetry, the diagonals or braces are alternately in tension and compression; similarly they are on the right but in the reverse order. Considering separately the ties and struts, we see that the internal

* This example is analogous to one of those studied by Maxwell in his memoir of 1870.

forces decrease from the ends towards the middle of the structure.

In this example again the diagrams have axes of symmetry.

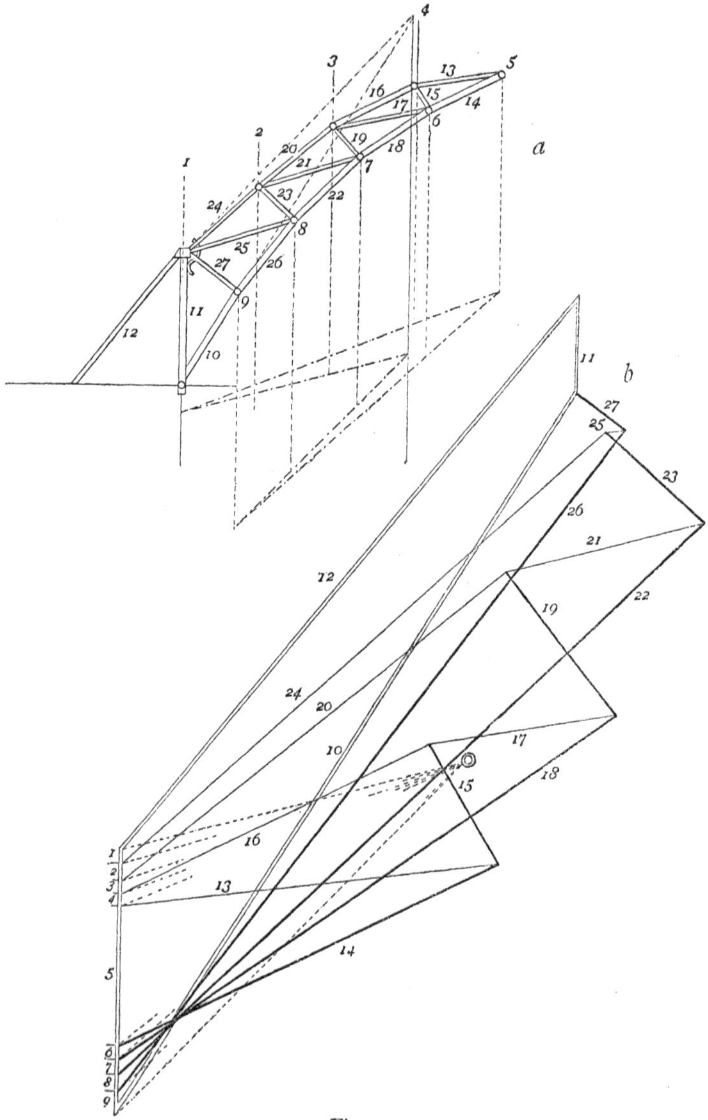

Fig. 12.

47. Diagram *a* of Fig. 12 represents a framed crane-post; the weight of the machine is distributed over the

different joints, and is represented by the sum of the forces **1, 2, 3, ..., 9**; the force **5** also includes the load the crane is required to lift.

All these forces are kept in equilibrium by the reactions at the supports, the magnitudes of which are obtained by resolving the resultant weight into three forces acting along the lines **10, 11, 12**. These forces, taken in the contrary sense, furnish the pressures which are supported by the strut **10**, and the column **11**, and the tension in the tie **12**.

48. These external forces may be determined as follows:

We take on the same vertical line segments representing the magnitudes of the forces 1, 2, 3 ... 9, and choose a pole arbitrarily; join the pole to the points (0, 1), (1, 2), (2, 3), ... (8, 9), (9, 0)*, and construct the corresponding funicular polygon. The vertical through the point where the extreme sides (**0**, **1**), (**9**, **0**) meet will be the line of action of the total weight of the crane and load, a weight represented in magnitude by a segment, which has the same initial point as the segment 1, and the same final point as the segment 9. If now we decompose the resultant weight, which is now known, into three components, whose lines of action are the straight lines **10, 11, 12**, employing the construction of Art. 40 (Fig. 6), we obtain the three forces 10, 11, 12. That is to say, these taken in the opposite sense and the given weights complete the system of external forces.

In order to obtain the diagram *b*, we construct first the polygon of the external forces, taking these forces in the order in which we encounter them in going round the contour of the structure. Then construct in succession the polygons corresponding to the joints: (**5, 13, 14**), (**4, 13, 15, 16**), (**6, 14, 15, 17, 18**), and so on in the manner just described.

The diagram thus obtained enables us to see that the bars of the upper part are in tension, and those of the lower part in compression; while the diagonals are alternately in tension and compression.

* Here (0, 1) represents the initial point of the segment 1, and (9, 0) the final point of segment 9. In the figure the rays, from the pole *O* and the sides of the corresponding funicular polygon, are shown by dotted lines

Select Works

PUBLISHED BY THE CLARENDON PRESS.

Aldis.

A TEXT-BOOK OF ALGEBRA; with Answers to the Examples. By W. STEADMAN ALDIS, M.A., Professor of Mathematics in Auckland College, New Zealand. Crown 8vo., *cloth,* 7*s.* 6*d.*

Baynes.

LESSONS ON THERMODYNAMICS. By R. E. BAYNES, M.A., Senior Student of Christ Church, Oxford, and Lee's Reader in Physics. Crown 8vo. *cloth,* 7*s.* 6*d.*

Cremona.

ELEMENTS OF PROJECTIVE GEOMETRY. By LUIGI CREMONA. Translated by C. LEUDESDORF, M.A. 8vo. 12*s.* 6*d.*

Donkin.

ACOUSTICS. By W. F. DONKIN, M.A., F.R.S., late Savilian Professor of Astronomy, Oxford. *Second Edition.* Crown 8vo. *cloth,* 7*s.* 6*d.*

Maxwell.

A TREATISE ON ELECTRICITY AND MAGNETISM. By J. CLERK MAXWELL, M.A., F.R.S. *Second Edition.* 2 vols. Demy 8vo. *cloth,* 1*l.* 11*s.* 6*d.*

ELEMENTARY TREATISE ON ELECTRICITY. By J. CLERK MAXWELL, M.A., F.R.S. Edited by WILLIAM GARNETT, M.A. Demy 8vo. *cloth,* 7*s.* 6*d.*

Minchin.

A TREATISE ON STATICS. By G. M. MINCHIN, M.A. *Third Edition, Corrected and Enlarged.*

VOL. I. Equilibrium of Coplanar Forces. 8vo. *cloth,* 9*s.*

VOL. II. Statics. 8vo. *cloth,* 16*s.*

UNIPLANAR KINEMATICS OF SOLIDS AND FLUIDS. By the same Author. Crown 8vo. *cloth,* 7*s.* 6*d.*

Nixon.

EUCLID REVISED. Containing the Essentials of the Elements of Plane Geometry as given by Euclid in his first Six Books. With numerous additional Propositions and Exercises. Edited by R. C. J. NIXON, M.A. *Second Edition.* Crown 8vo. 6*s.*

Sold separately as follows:—

Book I, 1*s.* Books I–IV, 3*s.*
Books I and II, 1*s.* 6*d.* Books V–VI, 3*s.*

EUCLID.—GEOMETRY IN SPACE. Containing parts of Euclid's Eleventh and Twelfth Books. By the same Editor. Crown 8vo. *cloth*, 3*s.* 6*d.*

Price.

TREATISE ON INFINITESIMAL CALCULUS. By BARTHOLOMEW PRICE, M.A., F.R.S., Professor of Natural Philosophy, Oxford.

VOL. I. Differential Calculus. *Second Edition.* 8vo. *cloth*, 14*s.* 6*d.*

VOL. II. Integral Calculus, Calculus of Variations, and Differential Equations. *Second Edition.* 8vo. *cloth*, 18*s.*

VOL. III. Statics, including Attractions; Dynamics of a Material Particle. *Second Edition.* 8vo. *cloth*, 16*s.*

VOL. IV. Dynamics of Material Systems; together with a Chapter on Theoretical Dynamics, by W. F. DONKIN, M.A., F.R.S. *Second Edition.* 8vo. *cloth*, 18*s.*

Stewart.

A TREATISE ON HEAT, with numerous Woodcuts and Diagrams. By BALFOUR STEWART, LL.D., F.R.S., Professor of Natural Philosophy in Owens College, Manchester. *Fifth Edition.* Extra fcap. 8vo. *cloth*, 7*s.* 6*d.*

Watson and Burbury.

THE MATHEMATICAL THEORY OF ELECTRICITY AND MAGNETISM. By H. W. WATSON, D.Sc., F.R.S., formerly Fellow of Trinity College, Cambridge; and S. H. BURBURY, M.A., formerly Fellow of St. John's College, Cambridge.

VOL. I. Electrostatics. 8vo. *cloth*, 10*s.* 6*d.*

VOL. II. Magnetism and Electrodynamics. 8vo. *cloth*, 10*s.* 6*d.*

A TREATISE ON THE APPLICATION OF GENERALISED COORDINATES TO THE KINETICS OF A MATERIAL SYSTEM. By the same Authors. 8vo. *cloth*, 6*s.*

Oxford
AT THE CLARENDON PRESS
LONDON: HENRY FROWDE
OXFORD UNIVERSITY PRESS WAREHOUSE, AMEN CORNER, E.C.

CLARENDON PRESS BOOKS

HISTORY

Greece, Italy, Egypt, etc

Clinton's Fasti Hellenici, from the LVIth to the CXXIIIrd Olympiad. Third edition. 4to. £1 14s. 6d. From the CXXIVth Olympiad to the Death of Augustus. Second edition. 4to. £1 12s. Epitome. 8vo. 6s. 6d.

Clinton's Fasti Romani, from the death of Augustus to the death of Heraclius. Two volumes. 4to. £2 2s. Epitome. 8vo. 7s.

Greswell's Fasti Temporis Catholici. 4 vols. 8vo. £2 10s. Tables and Introduction to Tables. 8vo. 15s. Origines Kalendariae Italicae. 4 vols. 8vo. £2 2s. Origines Kalendariae Hellenicae. 6 vols. 8vo. £4 4s.

A Manual of Greek Historical Inscriptions. By E. L. Hicks. New edition, revised by G. F. Hill. 8vo. 10s. 6d. net.

Latin Historical Inscriptions, illustrating the history of the Early Empire. By G. McN. Rushforth. 8vo. 10s. net.

Sources for Greek History between the Persian and Peloponnesian Wars. By G. F. Hill. 8vo. Reissue, revised. 10s. 6d. net.

Sources for Roman History, B.C. 133–70. By A. H. J. Greenidge and A. M. Clay. Crown 8vo. 5s. 6d. net.

A Manual of Ancient History. By G. Rawlinson. 2nd ed. 8vo. 14s.

Finlay's History of Greece from its Conquest by the Romans (B.C. 146) to A.D. 1864. A new edition, revised, and in part re-written, with many additions, by the Author, and edited by H. F. Tozer. 7 vols. 8vo. £3 10s.

The History of Sicily from the earliest times. By E. A. Freeman.
Vols. I and II. [The Native Nations: The Phoenician and Greek Settlements to the beginning of Athenian Intervention.] 8vo. £2 2s.
Vol. III. The Athenian and Carthaginian Invasions. £1 4s.
Vol. IV. From the Tyranny of Dionysios to the Death of Agathoklês. Edited from posthumous MSS, by A. J. Evans. £1 1s.

Italy and her Invaders (A.D. 376–814). With plates and maps. Eight volumes. 8vo. By T. Hodgkin. Vols. I–IV in the second edition.
I–II. The Visigothic, Hunnish, and Vandal Invasions, and the Herulian Mutiny. £2 2s.
III–IV. The Ostrogothic Invasion. The Imperial Restoration. £1 16s.
V–VI. The Lombard Invasion, and the Lombard Kingdom. £1 16s.
VII–VIII. Frankish Invasions, and the Frankish Empire. £1 4s.

The Dynasty of Theodosius; or, Seventy Years' Struggle with the Barbarians. By the same author. Crown 8vo. 6s.

Aetolia; its Geography, Topography, and Antiquities. By W. J. Woodhouse. With maps and illustrations. Royal 8vo. £1 1s. net.

The Islands of the Aegean. By H. F. Tozer. Crown 8vo. 8s. 6d.

Dalmatia, the Quarnero, and Istria; with Cettigne and Grado. By T. G. Jackson. Three volumes. With plates and illustrations. 8vo. £2 2s.

Cramer's Description of Asia Minor. Two volumes. 8vo. 11s.
Description of Ancient Greece. 3 vols. 8vo. 16s. 6d.

The Cities and Bishoprics of Phrygia. By W. M. RAMSAY. Royal 8vo. Vol. I, Part I. The Lycos Valley and South-Western Phrygia. 18s. net. Vol. I, Part II. West and West Central Phrygia. £1 1s. net.

Stories of the High Priests of Memphis, the Sethon of Herodotus, and the Demotic Tales of Khamnas. By F. LL. GRIFFITH. With Portfolio containing seven facsimiles. Royal 8vo. £2 7s. 6d. net.

The Arab Conquest of Egypt. By A. J. BUTLER. With maps and plans. 8vo. 16s. net.

Baghdad during the Abbasid Caliphate, from contemporary sources. By G. LE STRANGE. With eight plans. 8vo. 16s. net.

Archaeology

Ancient Khotan. Detailed report of Archaeological explorations in Chinese Turkestan carried out and described under the orders of H.M. Indian Government by M. AUREL STEIN. Vol. I. Text, with descriptive list of antiques, seventy-two illustrations in the text, and appendices. Vol. II. One hundred and nineteen collotype and other illustrations and a map. 2 vols. 4to. £5 5s. net.

Catalogue of the Coins in the Indian Museum, Calcutta, including the Cabinet of the Asiatic Society of Bengal. Vol. I. By VINCENT A. SMITH. Royal 8vo, 30s. net; or separately, Part I. The Early Foreign Dynasties and the Guptas, 15s. net. Part II. Ancient Coins of Indian Types, 6s. net. Part III. Persian, Mediaeval, South Indian, and Miscellaneous Coins, 10s. 6d. net. Vol. II. Part I. The Sultáns of Delhí. Part II. Contemporary Dynasties in India, by H. NELSON WRIGHT, with 25 plates. Royal 8vo. 30s. net. (The first section of Part II by Sir JAMES BOURDILLON.) (Published for the Trustees of the Indian Museum.)

Ancient Coptic Churches of Egypt. By A. J. BUTLER. 2 vols. 8vo. 30s.

A Catalogue of the Cyprus Museum. By J. L. MYRES and MAX OHNEFALSCH-RICHTER. 8vo. With eight plates, 7s. 6d. net.

A Catalogue of the Sparta Museum. By M. N. TOD and A. J. B. WACE. 8vo. 10s. 6d. net.

Catalogue of the Greek Vases in the Ashmolean Museum. By P. GARDNER. Small folio, linen, with 26 plates. £3 3s. net.

The Cults of the Greek States. By L. R. FARNELL. 8vo. Vols. I and II, with 61 plates and over 100 illustrations. £1 12s. net: Vols. III and IV, with 86 plates. £1 12s. net.

Classical Archaeology in Schools. By P. GARDNER and J. L. MYRES. 8vo. Second edition. Paper covers, 1s. net.

Introduction to Greek Sculpture. By L. E. UPCOTT. Second edition. Crown 8vo. 4s. 6d.

Marmora Oxoniensia, inscriptiones Graecae ad Chandleri exempla editae, cur. GUL. ROBERTS, 1791. Crown 8vo. 3s. 6d.

De Antiquis Marmoribus, Blasii Caryophili. 1828. 7s. 6d.

Fragmenta Herculanensia. A Catalogue of the Oxford copies of the Herculanean Rolls, with texts of several papyri. By W. SCOTT. Royal 8vo. £1 1s.

Thirty-six Engravings of Texts and Alphabets from the Herculanean Fragments. Folio. Small paper, 10s. 6d., large paper, £1 1s.

Herculanensium Voluminum Partes II. 1824. 8vo. 10s.

English History: Sources

Ancient Britain and the Invasions of Julius Caesar. By T. RICE HOLMES. 8vo. In the press.

Two of the Saxon Chronicles Parallel; with supplementary extracts from the others. A Revised Text, edited, with introduction, notes, appendices, and glossary, by C. PLUMMER and J. EARLE. Two volumes crown 8vo, leather-back. Vol. I. Text, appendices, and glossary. 10s. 6d. Vol. II. Introduction, notes, and index. 12s. 6d.

The Saxon Chronicles (787–1001 A. D.). Crown 8vo, stiff covers. 3s.

Baedae Opera Historica, edited by C. PLUMMER. Two volumes. Crown 8vo, leather back. £1 1s. net.

Handbook to the Land-Charters, and other Saxonic Documents, by J. EARLE. Crown 8vo. 16s.

The Crawford Collection of early Charters and Documents, now in the Bodleian Library. Edited by A. S. NAPIER and W. H. STEVENSON. Small 4to, cloth. 12s.

Asser's Life of Alfred, with the Annals of St. Neot, edited by W. H. STEVENSON. Crown 8vo. 12s. net.

The Alfred Jewel, an historical essay. With illustrations and a map, by J. EARLE. Small 4to, buckram. 12s. 6d. net.

Chronicles of London. Edited, with introduction and notes, by C. L. KINGSFORD. 8vo. 10s. 6d. net.

Dialogus de Scaccario (De necessariis observantiis Scaccarii dialogus) by Richard, Son of Nigel. Edited by A. HUGHES, C. G. CRUMP, and C. JOHNSON, with introduction and notes. 8vo. 12s. 6d. net.

The Song of Lewes. Edited from the MS, with introduction and notes, by C. L. KINGSFORD. Extra fcap 8vo. 5s.

Chronicon Galfridi le Baker de Swynebroke, edited by Sir E. MAUNDE THOMPSON, K.C.B. Small 4to, 18s.; cloth, gilt top, £1 1s.

Passio et Miracula Beati Olaui. Edited from the Twelfth-century MS by F. METCALFE. Small 4to. 6s.

Gascoigne's Theological Dictionary ('Liber Veritatum'): selected passages, illustrating the condition of Church and State, 1403–1458. With an introduction by J. E. THOROLD ROGERS. Small 4to. 10s. 6d.

Fortescue's Governance of England: otherwise called The Difference between an Absolute and a Limited Monarchy. A revised text, edited, with introduction, etc, by C. PLUMMER. 8vo, leather back. 12s. 6d.

The Protests of the Lords, including those which have been expunged, from 1624 to 1874; with historical introductions. By J. E. THOROLD ROGERS. In three volumes. 8vo. £2 2s.

Index to Wills proved in the Court of the Chancellor of the University of Oxford, etc. By J. GRIFFITHS. Royal 8vo. 3s. 6d.

The Clarendon Press Series of Charters, Statutes, etc

From the earliest times to 1307. By Bishop STUBBS.

Select Charters and other illustrations of English Constitutional History. Eighth edition. Crown 8vo. 8s. 6d.

From 1307 to 1558. In Preparation. By G. W. PROTHERO.

Select Statutes and other Constitutional Documents.

From 1558 to 1625.

Select Statutes and other Constitutional Documents of the Reigns of Elizabeth and James I. Third edition. Crown 8vo. 10s. 6d.

From 1625 to 1660. By S. R. GARDINER.

The Constitutional Documents of the Puritan Revolution. Third edition. Crown 8vo. 10s. 6d.

Calendars, etc

Calendar of Charters and Rolls preserved in the Bodleian Library. 8vo. £1 11s. 6d.

Calendar of the Clarendon State Papers preserved in the Bodleian Library. In three volumes. 1869–76.

Vol. I. From 1523 to January 1649. 8vo. 18s. Vol. II. From 1649 to 1654. 8vo. 16s. Vol. III. From 1655 to 1657. 8vo. 14s.

Hakluyt's Principal Navigations, being narratives of the Voyages of the Elizabethan Seamen to America. Selection edited by E. J. PAYNE. Crown 8vo, with portraits. Second edition. First and Second Series, 5s. each.

Also abridged, in one volume, with additional notes, maps, &c., by C. RAYMOND BEAZLEY. Crown 8vo, with illustrations. 4s. 6d. Also, separately, 'The Voyages of Hawkins, Frobisher, and Drake.' 2s. 6d.

Aubrey's 'Brief Lives,' set down between the Years 1669 and 1696. Edited from the Author's MSS by A. CLARK. Two volumes. 8vo. £1 5s.

Whitelock's Memorials of English Affairs from 1625 to 1660. 4 vols. 8vo. £1 10s.

Ludlow's Memoirs, 1625–1672. Edited, with Appendices of Letters and illustrative documents, by C. H. FIRTH. Two volumes. 8vo. £1 16s.

Luttrell's Diary. A brief Historical Relation of State Affairs, 1678–1714. Six volumes. 8vo. £1 4s.

Burnet's History of James II. 8vo. 9s. 6d.

Life of Sir M. Hale, with **Fell's Life of Dr. Hammond.** Small 8vo. 2s. 6d.

ENGLISH HISTORY

Burnet's History of My Own Time. A new edition based on that of M. J. Routh. Edited by Osmund Airy. Vol. I. 12s. 6d. Vol. II. (Completing Charles the Second, with Index to Vols. I and II.) 12s. 6d.

Supplement, derived from Burnet's Memoirs, Autobiography, etc., all hitherto unpublished. Edited by H. C. Foxcroft, 1902. 8vo. 16s. net.

Carte's Life of James Duke of Ormond. A new edition carefully compared with the original MSS. Six volumes. 8vo. £1 5s.

The Whitefoord Papers, from 1739 to 1810. Edited by W. A. S. Hewins. 8vo. 12s. 6d.

History of Oxford

A complete list of the Publications of the Oxford Historical Society can be obtained from Mr. Frowde.

Manuscript Materials relating to the History of Oxford; contained in the printed catalogues of the Bodleian and College Libraries. By F. Madan. 8vo. 7s. 6d.

The Early Oxford Press. A Bibliography of Printing and Publishing at Oxford, '1468'-1640. With notes, appendices, and illustrations. By F. Madan. 8vo. 18s.

Bibliography

Cotton's Typographical Gazetteer. First Series. 8vo. 12s. 6d.

Ebert's Bibliographical Dictionary. 4 vols. 8vo. £3 3s. net.

Bishop Stubbs's and Professor Freeman's Books

The Constitutional History of England, in its Origin and Development. By W. Stubbs. Library edition. Three volumes. Demy 8vo. £2 8s. Also in three volumes, crown 8vo, price 12s. each.

Seventeen Lectures on the Study of Mediaeval and Modern History and kindred subjects, 1867–1884. By the same. Third edition, revised and enlarged, 1900. Crown 8vo, half-roan. 8s. 6d.

History of the Norman Conquest of England; its Causes and Results. By E. A. Freeman. Vols. I, II and V (English edition) are out of print.

Vols. III and IV. £1 1s. each. Vol. VI (Index). 10s. 6d.

A Short History of the Norman Conquest of England. Third edition. By the same. Extra fcap 8vo. 2s. 6d.

The Reign of William Rufus and the Accession of Henry the First. By the same. Two volumes. 8vo. £1 16s.

Companion to English History (Middle Ages). Edited by F. P. Barnard. With 97 illustrations. Crown 8vo. 8s. 6d. net.

School History of England to the death of Victoria. With maps, plans, etc. By O. M. Edwards, R. S. Rait and others. Crown 8vo, 3s. 6d.

Special Periods and Biographies

Life and Times of Alfred the Great, being the Ford Lectures for 1901. By C. Plummer. 8vo. 5s. net.

The Domesday Boroughs. By Adolphus Ballard. 8vo, with four plans. 6s. 6d. net.

Villainage in England. Essays in English Mediaeval History. By P. Vinogradoff. 8vo, leather back. 16s.

English Society in the Eleventh Century. Essays in English Mediaeval History. By P. Vinogradoff. 8vo. In the press.

The Gild Merchant: a contribution to British municipal history. By C. Gross. Two volumes. 8vo, leather back, £1 4s.

The Welsh Wars of Edward I; a contribution to mediaeval military history. By J. E. Morris. 8vo. 9s. 6d. net.

The Great Revolt of 1381. By C. Oman. With two maps. 8vo. 8s. 6d. net.

Lancaster and York. A Century of English History (A. D. 1399–1485). By Sir J. H. Ramsay. Two volumes. 8vo, with Index, £1 17s. 6d. Index separately, 1s. 6d.

Life and Letters of Thomas Cromwell. By R. B. Merriman. In two volumes. [Vol. I, Life and Letters, 1523–1535, etc. Vol. II, Letters, 1536–1540, notes, index, etc.] 8vo. 18s. net.

A History of England, principally in the Seventeenth Century. By L. von Ranke. Translated under the superintendence of G. W. Kitchin and C. W. Boase. Six volumes. 8vo. £3 3s. Index separately, 1s.

Sir Walter Ralegh, a Biography, by W. Stebbing. Post 8vo. 6s. net.

Biographical Memoir of Dr. William Markham, Archbishop of York, by his great-grandson, Sir Clements Markham, K.C.B. 8vo. With photogravure portrait. 5s. net.

The Life and Works of John Arbuthnot. By G. A. Aitken. 8vo, cloth extra, with Portrait. 16s.

Life and Letters of Sir Henry Wotton. By L. Pearsall-Smith. 8vo. Two volumes. 25s. net.

Great Britain and Hanover. By A. W. Ward. Crown 8vo. 5s.

History of the Peninsular War. By C. Oman. To be completed in six volumes, 8vo, with many maps, plans, and portraits.

Already published: Vol. I. 1807–1809, to Corunna. 14s. net. Vol. II. 1809, to Talavera. 14s. net. Vol. III. In the Press.

Anglo-Chinese Commerce and Diplomacy: mainly in the nineteenth century. By A. J. Sargent. In the press.

Frederick York Powell. A Life and a selection from his Letters and Occasional Writings. By Oliver Elton. Two volumes. 8vo. With photogravure portraits, facsimiles, etc. 21s. net.

David Binning Monro: a Short Memoir. By J. Cook Wilson. 8vo, stiff boards, with portrait. 2s. net.

F. W. Maitland. Two lectures by A. L. Smith. 8vo. In the press.

History and Geography of America and the British Colonies

For other Geographical books, see page 10.

History of the New World called America. By E. J. Payne.
Vol. I. 8vo. 18s. Bk. I. The Discovery. Bk. II, Part I. Aboriginal America.
Vol. II. 8vo. 14s. Bk. II, Part II. Aboriginal America (concluded).

The Canadian War of 1812. By C. P. Lucas, C.B. 8vo. With eight maps. 12s. 6d. net.

Historical Geography of the British Colonies. By C. P. Lucas, C.B. Crown 8vo.

Introduction. New edition by H. E. Egerton. 1903. With eight maps. 3s. 6d. In cheaper binding, 2s. 6d.

Vol. I. The Mediterranean and Eastern Colonies. With 13 maps. Second edition, revised and brought up to date, by R. E. Stubbs. 1906. 5s.

Vol. II. The West Indian Colonies. With twelve maps. Second edition, revised and brought up to date, by C. Atchley, I.S.O. 1905. 7s. 6d.

Vol. III. West Africa. Second Edition. Revised to the end of 1899 by H. E. Egerton. With five maps. 7s. 6d.

Vol. IV. South and East Africa. Historical and Geographical. With eleven maps. 9s. 6d.
Also Part I. Historical. 1898. 6s. 6d. Part II. 1903. Geographical. 3s. 6d.

Vol. V. Canada, Part I. 1901. 6s.

Vol. VI. Australasia. By J. D. Rogers. 1907. With 22 maps. 7s. 6d. Also Part I, Historical, 4s. 6d. Part II, Geographical, 3s. 6d.

History of the Dominion of Canada. By W. P. Greswell. Crown 8vo. 7s. 6d.

Geography of the Dominion of Canada and Newfoundland. By the same author. With ten maps. 1891. Crown 8vo. 6s.

Geography of Africa South of the Zambesi. With maps. 1892. By the same author. Crown 8vo. 7s. 6d.

The Claims of the Study of Colonial History upon the attention of the University of Oxford. An inaugural lecture delivered on April 28, 1906, by H. E. Egerton. 8vo, paper covers, 1s. net.

Historical Atlas. Europe and her Colonies, 27 maps. 35s. net.

Cornewall-Lewis's Essay on the Government of Dependencies. Edited by C. P. Lucas, C.B. 8vo, quarter-bound, 14s.

History of India

The Imperial Gazetteer of India. New edition. To be completed in twenty-six volumes. 8vo. Subscription price, cloth, £5 net; morocco back, £6 6s. net. The four volumes of 'The Indian Empire' (I, III, IV are ready) separately 6s. net each, in cloth, or 7s. 6d. net with morocco back; the Atlas separately 15s. net in cloth, or 17s. 6d. net with morocco back. Subscriptions may be sent through any bookseller.

Reprints from the Imperial Gazetteer.

A sketch of the Flora of British India. By Sir JOSEPH HOOKER. 8vo. Paper covers. 1s. net.

The Indian Army. A sketch of its History and Organization. 8vo. Paper covers. 1s. net.

A Brief History of the Indian Peoples. By Sir W. W. HUNTER. Revised up to 1903 by W. H. HUTTON. Eighty-ninth thousand. 3s. 6d.

Rulers of India. Edited by Sir W. W. HUNTER. Crown 8vo. 2s. 6d. each.

Bábar. By S. LANE-POOLE.
Akbar. By Colonel MALLESON.
Albuquerque. By H. MORSE STEPHENS.
Aurangzíb. By S. LANE-POOLE.
Mádhava Ráo Sindhia. By H. G. KEENE.
Lord Clive. By Colonel MALLESON.
Dupleix. By Colonel MALLESON.
Warren Hastings. By Captain L. J. TROTTER.
The Marquis of Cornwallis. By W. S. SETON-KARR.
Haidar Alí and Tipú Sultán. By L. B. BOWRING.
The Marquis Wellesley, K.G. By W. H. HUTTON.
Marquess of Hastings. By Major ROSS-OF-BLADENSBURG.
Mountstuart Elphinstone. By J. S. COTTON.
Sir Thomas Munro. By J. BRADSHAW.
Earl Amherst. By ANNE T. RITCHIE and R. EVANS.
Lord William Bentinck. By D. C. BOULGER.
The Earl of Auckland. By Captain L. J. TROTTER.
Viscount Hardinge. By his son, Viscount HARDINGE.
Ranjit Singh. By Sir L. GRIFFIN.
The Marquess of Dalhousie. By Sir W. W. HUNTER.
John Russell Colvin. By Sir A. COLVIN.
Clyde and Strathnairn. By Major-General Sir O. T. BURNE.
Earl Canning. By Sir H. S. CUNNINGHAM.
Lord Lawrence. By Sir C. AITCHISON.
The Earl of Mayo. By Sir W. W. HUNTER.

Supplementary volumes.

Asoka. By V. A. SMITH. 3s. 6d.
James Thomason. By Sir R. TEMPLE. 3s. 6d.
Sir Henry Lawrence, the Pacificator. By Lieut.-General J. J. McLEOD INNES. 3s. 6d.

HISTORY OF INDIA

The Government of India, being a digest of the Statute Law relating thereto; with historical introduction and illustrative documents. By Sir C. P. Ilbert. Second edition, 1907. 10s. 6d. net.

The Early History of India from 600 B.C. to the Muhammadan Conquest, including the invasion of Alexander the Great. By V. A. Smith. 8vo. With maps, plans, and other illustrations. 14s. net.

The English Factories in India, 1618–1621. By W. Foster. 8vo. (Published under the patronage of His Majesty's Secretary of State for India in Council.) 12s. 6d. net.

Court Minutes of the East India Company, 1635–1639. Edited by W. Foster. 8vo. In the press.

Wellesley's Despatches, Treaties, and other Papers relating to his Government of India. Selection edited by S. J. Owen. 8vo. £1 4s.

Wellington's Despatches, Treaties, and other Papers relating to India. Selection edited by S. J. Owen. 8vo. £1 4s.

Hastings and the Rohilla War. By Sir J. Strachey. 8vo. 10s. 6d.

European History

Historical Atlas of Modern Europe, from the Decline of the Roman Empire. Containing 90 maps, with letterpress to each map: the maps printed by W. & A. K. Johnston, Ltd., and the whole edited by R. L. Poole.

In one volume, imperial 4to, half-persian, £5 15s. 6d. net; or in selected sets—British Empire, etc, at various prices from 30s. to 35s. net each; or in single maps, 1s. 6d. net each. Prospectus on application.

Genealogical Tables illustrative of Modern History. By H. B. George. Fourth (1904) edition. Oblong 4to, boards. 7s. 6d.

The Life and Times of James the First of Aragon. By F. D. Swift. 8vo. 12s. 6d.

A History of France. With maps, plans, and Tables. By G. W. Kitchin. New edition. In three volumes, crown 8vo, each 10s. 6d.

Vol. I, to 1453. Vol. II, 1453–1624. Vol. III, 1624–1793.

The Principal Speeches of the Statesmen and Orators of the French Revolution, 1789–1795. With introductions, notes, etc. By H. Morse Stephens. Two volumes. Crown 8vo. £1 1s.

Napoleonic Statesmanship: Germany. By H. A. L. Fisher. 8vo, with maps. 12s. 6d. net.

De Tocqueville's L'Ancien Régime et la Révolution. Edited, with introductions and notes, by G. W. Headlam. Crown 8vo. 6s.

Documents of the French Revolution, 1789–1791. By L. G. Wickham Legg. Crown 8vo. Two volumes. 12s. net.

Thiers' Moscow Expedition, edited, with introductions and notes, by H. B. George. Crown 8vo, with 6 maps. 5s.

Geography and Anthropology

Relations of Geography and History. By H. B. George. With two maps. Crown 8vo. Third edition. 4s. 6d.

The Dawn of Modern Geography. By C. R. Beazley. In three volumes. Vol. I (to A.D. 900). Vol. II (A.D. 900-1260). 15s. net each. Vol. III. 20s. net.

Regions of the World. Geographical Memoirs under the general editorship of H. J. Mackinder. Large 8vo. Each volume contains maps and diagrams. 7s. 6d. net per volume.

Britain and the British Seas. Second edition. By H. J. Mackinder. Central Europe. By John Partsch. The Nearer East. By D. G. Hogarth. North America. By J. Russell. India. By Sir Thomas Holdich. The Far East. By Archibald Little.

The Face of the Earth (Das Antlitz der Erde). By Eduard Suess. Translated by Hertha Sollas. Vols. I, II. 25s. net each.

The Oxford Geographies. By A. J. Herbertson. Crown 8vo.
Vol. I. The Preliminary Geography, Ed. 2, 72 maps and diagrams, 1s. 6d.
Vol. II. The Junior Geography, Ed. 2, 166 maps and diagrams, 2s.
Vol. III. The Senior Geography, Ed. 2, with 117 maps and diagrams, 2s. 6d.

Geography for Schools, by A. Hughes. Crown 8vo. 2s. 6d.

Anthropological Essays presented to Edward Burnett Tylor in honour of his seventy-fifth birthday; by H. Balfour, A. E. Crawley, D. J. Cunningham, L. R. Farnell, J. G. Frazer, A. C. Haddon, E. S. Hartland, A. Lang, R. R. Marett, C. S. Myers, J. L. Myres, C. H. Read, Sir J. Rhŷs, W. Ridgeway, W. H. R. Rivers, C. G. Seligmann, T. A. Joyce, N. W. Thomas, A. Thomson, E. Westermarck; with a bibliography by Barbara W. Freire-Marreco. Imperial 8vo. 15s. net.

The Evolution of Culture, and other Essays, by the late Lieut.-Gen. A. Lane-Fox Pitt-Rivers; edited by J. L. Myres, with an Introduction by H. Balfour. 8vo, with 21 plates, 7s. 6d. net.

Dubois' Hindu Manners, Customs, and Ceremonies. Translated and edited with notes, corrections, and biography, by H. K. Beauchamp. Third edition. Crown 8vo. 6s. net. On India Paper, 7s. 6d. net.

The Melanesians, studies in their Anthropology and Folk-Lore. By R. H. Codrington. 8vo. 16s.

Iceland and the Faroes. By N. Annandale. With twenty-four illustrations and an appendix on the Celtic Pony, by F. H. A. Marshall. Crown 8vo. 4s. 6d. net.

The Masai, their Language and Folk-lore. By A. C. Hollis. With introduction by Sir Charles Eliot. 8vo. With 27 full-page illustrations and a map. 14s. net.

Celtic Folklore: Welsh and Manx. By J. Rhŷs. Two volumes. 8vo. £1 1s.

Studies in the Arthurian Legend. By J. Rhŷs. 8vo. 12s. 6d.

The Mediaeval Stage, from classical times through folk-play and minstrelsy to Elizabethan drama. By E. K. Chambers. With two illustrations. 8vo. £1 5s. net.

PHILOSOPHY

Modern Philosophy

Bacon's Novum Organum, edited, with introduction, notes, etc, by T. Fowler. Second edition. 8vo. 15s.

Novum Organum, edited, with notes, by G. W. Kitchin. 8vo. 9s. 6d.

Bentham's Introduction to the Principles of Morals and Legislation. Crown 8vo. 6s. 6d.

The Works of George Berkeley, formerly Bishop of Cloyne. With prefaces, annotations, appendices, and an account of his Life and Philosophy, by A. C. Fraser. New edition (1901) in crown 8vo. Four volumes. £1 4s.

Some copies of the 8vo edition of the *Life* are still on sale, price 16s.

Selections from Berkeley, with introduction and notes, for the use of Students. By the same Editor. Fifth edition. Crown 8vo. 7s. 6d.

The Cambridge Platonists: being selections from the Writings of Benjamin Whichcote, John Smith, and Nathanael Culverwel, with introduction by E. T. Campagnac. Crown 8vo. 6s. 6d. net.

Leibniz's Monadology and other Philosophical Writings, translated, with introduction and notes, by R. Latta. Crown 8vo. 8s. 6d.

Locke's Essay concerning Human Understanding. Collated and annotated with prolegomena, biographical, critical, and historical, by A. C. Fraser. Two volumes. 8vo. £1 12s.

Locke's Conduct of the Understanding. Edited by T. Fowler. Extra fcap 8vo. 2s. 6d.

A Study in the Ethics of Spinoza. By H. H. Joachim. 8vo. 10s. 6d. net.

Hume's Treatise on Human Nature, reprinted from the original edition in three volumes, and edited by L. A. Selby-Bigge. Second edition. Crown 8vo. 6s. net.

Hume's Enquiry concerning the Human Understanding, and an Enquiry concerning the Principles of Morals. Edited by L. A. Selby-Bigge. Crown 8vo. Second edition. 6s. net.

British Moralists, being Selections from writers principally of the eighteenth century. Edited by L. A. Selby-Bigge. Two volumes. Crown 8vo. 12s. net. Uniform with Hume's Treatise and Enquiry, and Berkeley's Works.

Butler's Works, edited by W. E. Gladstone. Two volumes. Medium 8vo, 14s. each, or Crown 8vo, 10s. 6d. Also, separately—Vol. I (Analogy), 5s. 6d. Vol. II (Sermons), 5s.

Recent Philosophy

The Logic of Hegel, translated from the Encyclopaedia of the Philosophical Sciences, with Prolegomena, by W. WALLACE. Second edition. Two volumes. Crown 8vo. 10s. 6d. each.

Hegel's Philosophy of Mind, translated from Encyclopaedia of Philosophical Sciences, with five introductory essays, by W. WALLACE. Crown 8vo. 10s. 6d.

Lotze's Logic, in Three Books—of Thought, of Investigation, and of Knowledge. Translated by B. BOSANQUET. Second ed. 2 vols. Cr. 8vo. 12s.

Lotze's Metaphysic, in Three Books—Ontology, Cosmology, and Psychology. Translated by B. BOSANQUET. Second ed. 2 vols. Cr. 8vo. 12s.

Bluntschli's Theory of the State. Translated from the sixth German edition. Third edition, 1901. Crown 8vo, half-bound, 8s. 6d.

Green's Prolegomena to Ethics. Edited by A. C. BRADLEY. Fifth edition, 1906. With a Preface by E. CAIRD. Crown 8vo. 6s. net.

Types of Ethical Theory, by J. MARTINEAU. Third edition. Two volumes. Crown 8vo. 15s.

A Study of Religion: its Sources and Contents. By the same author. Second edition. Two volumes. Crown 8vo. 15s.

The Principles of Morals. By T. FOWLER and J. M. WILSON. 8vo. 14s. Also, separately—Part I, 3s. 6d. Part II, 10s. 6d.

Logic; or, The Morphology of Knowledge. By B. BOSANQUET. Two volumes. 8vo. £1 1s.

Lectures and Essays on Natural Theology and Ethics. By W. WALLACE. Edited, with biographical introduction, by E. CAIRD. With portrait. 8vo. 12s. 6d.

Studies in History and Jurisprudence. By Rt. Hon. J. BRYCE. 1901. 2 vols. 8vo. £1 5s. net.

The Theory of Good and Evil. By H. RASHDALL. 8vo. 2 vols. 14s. net.

The Herbert Spencer Lectures. 1905, by FREDERIC HARRISON. 8vo, paper covers, 2s. net. 1907. Probability, the Foundation of Eugenics. By FRANCIS GALTON. 8vo. 1s. net.

An Introduction to Logic. By H. W. B. JOSEPH. 8vo. 9s. 6d. net.

Essay on Truth. By H. H. JOACHIM. 8vo. 6s. net.

Elementary Logic

The Elements of Deductive Logic. By T. FOWLER. Tenth edition, with a collection of examples. Extra fcap 8vo. 3s. 6d.

The Elements of Inductive Logic. By the same. Sixth edition. Extra fcap 8vo. 6s. In one volume with Deductive Logic, 7s. 6d.

LAW

Jurisprudence

Bentham's Fragment on Government. Edited by F. C. MONTAGUE. 8vo. 7s. 6d.

Bentham's Introduction to the Principles of Morals and Legislation. Second edition. Crown 8vo. 6s. 6d.

Studies in History and Jurisprudence. By the Right Hon. JAMES BRYCE. 1901. Two volumes. 8vo. £1 5s. net.

The Elements of Jurisprudence. By T. E. HOLLAND. Tenth edition. 1906. 8vo. 10s. 6d.

Elements of Law, considered with reference to Principles of General Jurisprudence. By Sir W. MARKBY, K.C.I.E. Sixth edition revised, 1905. 8vo. 12s. 6d.

Roman Law

Imperatoris Iustiniani Institutionum Libri Quattuor; with introductions, commentary, and translation, by J. B. MOYLE. Two volumes. 8vo. Vol. I (fourth edition, 1903), 16s.; Vol. II, Translation (fourth edition, 1906), 6s.

The Institutes of Justinian, edited as a recension of the Institutes of Gaius. By T. E. HOLLAND. Second edition. Extra fcap 8vo. 5s.

Select Titles from the Digest of Justinian. By T. E. HOLLAND and C. L. SHADWELL. 8vo. 14s.

Also, sold in parts, in paper covers: Part I. Introductory Titles. 2s. 6d. Part II. Family Law. 1s. Part III. Property Law. 2s. 6d. Part IV. Law of Obligations. No. 1. 3s. 6d. No. 2. 4s. 6d.

Gai Institutionum Iuris Civilis Commentarii Quattuor: with a translation and commentary by the late E. POSTE. Fourth edition. Revised and enlarged, by E. A. WHITTUCK, with an historical introduction by A. H. J. GREENIDGE. 8vo. 16s. net.

Institutes of Roman Law, by R. SOHM. Translated by J. C. LEDLIE: with an introductory essay by E. GRUEBER. Third edition. 8vo. 16s. net.

Infamia; its place in Roman Public and Private Law. By A. H. J. GREENIDGE. 8vo. 10s. 6d.

Legal Procedure in Cicero's Time. By A. H. J. GREENIDGE. 8vo. £1 1s.

The Roman Law of Damage to Property: being a commentary on the title of the Digest 'Ad Legem Aquiliam' (ix. 2), with an introduction to the study of the Corpus Iuris Civilis. By E. GRUEBER. 8vo. 10s. 6d.

Contract of Sale in the Civil Law. By J. B. MOYLE. 8vo. 10s. 6d.

The Principles of German Civil Law. By ERNEST J. SCHUSTER. 8vo. 12s. 6d. net.

English Law

Principles of the English Law of Contract, and of Agency in its relation to Contract. By Sir W. R. Anson. Eleventh edition. 1906. 8vo. 10s. 6d.

Law and Custom of the Constitution. By the same. In two parts.

Part I. Parliament. Third edition. 8vo. 12s. 6d. (Out of print.)
Part II. The Crown. Third edition in preparation.

Calendar of Charters and Rolls, containing those preserved in the Bodleian Library. 8vo. £1 11s. 6d.

Introduction to the History of the Law of Real Property. By Sir K. E. Digby. Fifth edition. 8vo. 12s. 6d.

Handbook to the Land-Charters, and other Saxonic Documents. By J. Earle. Crown 8vo. 16s.

Fortescue's Difference between an Absolute and a Limited Monarchy. Text revised and edited, with introduction, etc, by C. Plummer. 8vo, leather back, 12s. 6d.

Legislative Methods and Forms. By Sir C. P. Ilbert, K.C.S.I. 1901. 8vo, leather back, 16s.

Modern Land Law. By E. Jenks. 8vo. 15s.

Essay on Possession in the Common Law. By Sir F. Pollock and Sir R. S. Wright. 8vo. 8s. 6d.

Outline of the Law of Property. By T. Raleigh. 8vo. 7s. 6d.

Villainage in England. By P. Vinogradoff. 8vo, leather back, 16s.

Law in Daily Life. By Rud. von Jhering. Translated with Notes and Additions by H. Goudy. Crown 8vo. 3s. 6d. net.

Cases illustrating the Principles of the Law of Torts, with table of all Cases cited. By F. R. Y. Radcliffe and J. C. Miles. 8vo. 1904. 12s. 6d. net.

Constitutional Documents

Select Charters and other Illustrations of English Constitutional History, from the earliest times to Edward I. Arranged and edited by W. Stubbs. Eighth edition. 1900. Crown 8vo. 8s. 6d.

Select Statutes and other Constitutional Documents, illustrative of the reigns of Elizabeth and James I. Edited by G. W. Prothero. Third edition. Crown 8vo. 10s. 6d.

Constitutional Documents of the Puritan Revolution, selected and edited by S. R. Gardiner. Third edition. Crown 8vo. 10s. 6d.

International Law

International Law. By W. E. Hall. Fifth edition by J. B. Atlay. 1904. 8vo. £1 1s. net.

Treatise on the Foreign Powers and Jurisdiction of the British Crown. By W. E. Hall. 8vo. 10s. 6d.

The European Concert in the Eastern Question, a collection of treaties and other public acts. Edited, with introductions and notes, by T. E. Holland. 8vo. 12s. 6d.

Studies in International Law. By T. E. Holland. 8vo. 10s. 6d.

Gentilis Alberici de Iure Belli Libri Tres edidit T. E. Holland. Small quarto, half-morocco. £1 1s.

The Law of Nations. By Sir T. Twiss. Part I. In time of peace. New edition, revised and enlarged. 8vo. 15s.

Colonial and Indian Law

The Government of India, being a Digest of the Statute Law relating thereto, with historical introduction and illustrative documents. By Sir C. P. Ilbert, K.C.S.I. Second edition. 8vo, cloth. 10s. 6d. net.

British Rule and Jurisdiction beyond the Seas. By the late Sir H. Jenkyns, K.C.B., with a preface by Sir C. P. Ilbert, and a portrait of the author. 1902. 8vo, leather back, 15s. net.

Cornewall-Lewis's Essay on the Government of Dependencies. Edited by C. P. Lucas, C.B. 8vo, leather back, 14s.

An Introduction to Hindu and Mahommedan Law for the use of students. 1906. By Sir W. Markby, K.C.I.E. 6s. net.

Land-Revenue and Tenure in British India. By B. H. Baden-Powell, C.I.E. With map. Second edition, revised by T. W. Holderness, C.S.I. (1907.) Crown 8vo. 5s. net.

Land-Systems of British India, being a manual of the Land-Tenures, and of the systems of Land-Revenue administration. By the same. Three volumes. 8vo, with map. £3 3s.

Anglo-Indian Codes, by Whitley Stokes. 8vo.
Vol. I. Substantive Law. £1 10s. Vol. II. Adjective Law. £1 15s.
1st supplement, 2s. 6d. 2nd supplement, to 1891, 4s. 6d. In one vol., 6s. 6d.

The Indian Evidence Act, with notes by Sir W. Markby, K.C.I.E. 8vo. 3s. 6d. net (published by Mr. Frowde).

Corps de Droit Ottoman : un Recueil des Codes, Lois, Règlements, Ordonnances et Actes les plus importants du Droit Intérieur, et d'Études sur le Droit Coutumier de l'Empire Ottoman. Par George Young. Seven vols. 8vo. Part I (Vols. I-III), cloth, £2 17s. 6d. net; paper covers, £2 12s. 6d. net: Part II (Vols. IV-VII), cloth, £1 17s. net, paper covers, £1 11s. 6d. net. Parts I and II can be obtained separately, but the price of either Part, bought alone, will be £2 12s. 6d. net in paper covers, or £2 17s. 6d. net in cloth.

Political Science and Economy

For Bryce's *Studies* and other books on general jurisprudence and political science, see p. 13.

Industrial Organization in the 16th and 17th Centuries. By G. Unwin. 8vo. 7s. 6d. net.

Relations of the Advanced and Backward Races of Mankind, the Romanes Lecture for 1902. By J. Bryce. 8vo. 2s. net.

Cornewall-Lewis's Remarks on the Use and Abuse of some Political Terms. New edition, with introduction by T. Raleigh. Crown 8vo, paper, 3s. 6d.; cloth, 4s. 6d.

Adam Smith's Wealth of Nations. Edited by J. E. Thorold Rogers. Two volumes. 8vo. £1 1s. net.

Adam Smith's Lectures on Justice, Police, Revenue and Arms. Edited with introduction and notes by E. Cannan. 8vo. 10s. 6d. net.

Bluntschli's Theory of the State. Translated from the sixth German edition. Third edition. 1901. Crown 8vo, half-bound, 8s. 6d.

Co-operative Production. By B. Jones. With preface by A. H. Dyke-Acland. Two volumes. Crown 8vo. 15s.

Elementary Political Economy. By E. Cannan. Fourth edition. Extra fcap 8vo, 1s.

Elementary Politics. By T. Raleigh. Sixth edition revised. Extra fcap 8vo, stiff covers, 1s.

A Geometrical Political Economy. Being an elementary Treatise on the method of explaining some Theories of Pure Economic Science by diagrams. By H. Cunynghame, C.B. Cr. 8vo. 2s. 6d. net.

The Elements of Railway Economics. By W. M. Acworth. Crown 8vo. Second impression. 2s. net.

Economic Documents

Ricardo's Letters to Malthus (1810-1823). Edited by J. Bonar. 8vo. 7s. 6d.

Letters to Trower and others (1811-1823). Edited by J. Bonar and J. H. Hollander. 8vo. 7s. 6d.

Lloyd's Prices of Corn in Oxford, 1583-1830. 8vo. 1s.

The History of Agriculture and Prices in England, A.D. 1259-1793. By J. E. Thorold Rogers.

Vols. I and II (1259-1400). 8vo. £2 2s.
Vols. III and IV (1401-1582). 8vo. £2 10s.
Vols. V and VI (1583-1702). 8vo. £2 10s.
Vol. VII. In two Parts (1702-1793). 8vo. £2 10s.

First Nine Years of the Bank of England. By the same. 8vo. 8s. 6d.

www.ingramcontent.com/pod-product-compliance
Lightning Source LLC
LaVergne TN
LVHW101440200926
848483LV00004B/414